Springer Theses

Recognizing Outstanding Ph.D. Research

Aims and Scope

The series "Springer Theses" brings together a selection of the very best Ph.D. theses from around the world and across the physical sciences. Nominated and endorsed by two recognized specialists, each published volume has been selected for its scientific excellence and the high impact of its contents for the pertinent field of research. For greater accessibility to non-specialists, the published versions include an extended introduction, as well as a foreword by the student's supervisor explaining the special relevance of the work for the field. As a whole, the series will provide a valuable resource both for newcomers to the research fields described, and for other scientists seeking detailed background information on special questions. Finally, it provides an accredited documentation of the valuable contributions made by today's younger generation of scientists.

Theses may be nominated for publication in this series by heads of department at internationally leading universities or institutes and should fulfill all of the following criteria

- They must be written in good English.
- The topic should fall within the confines of Chemistry, Physics, Earth Sciences, Engineering and related interdisciplinary fields such as Materials, Nanoscience, Chemical Engineering, Complex Systems and Biophysics.
- The work reported in the thesis must represent a significant scientific advance.
- If the thesis includes previously published material, permission to reproduce this must be gained from the respective copyright holder (a maximum 30% of the thesis should be a verbatim reproduction from the author's previous publications).
- They must have been examined and passed during the 12 months prior to nomination.
- Each thesis should include a foreword by the supervisor outlining the significance of its content.
- The theses should have a clearly defined structure including an introduction accessible to new PhD students and scientists not expert in the relevant field.

Indexed by zbMATH.

More information about this series at http://www.springer.com/series/8790

Servet Ozdemir

Electronic Properties of Rhombohedral Graphite

Doctoral Thesis accepted by The University of Manchester, UK

Springer

Author
Dr. Servet Ozdemir
Institute of Photonic and Quantum Sciences
Heriot-Watt University
Edinburgh, UK

Supervisor
Prof. K. S. Novoselov
Centre for Advanced 2D Materials
National University of Singapore
Singapore, Singapore

ISSN 2190-5053 ISSN 2190-5061 (electronic)
Springer Theses
ISBN 978-3-030-88306-5 ISBN 978-3-030-88307-2 (eBook)
https://doi.org/10.1007/978-3-030-88307-2

This Springer imprint is published by the registered company Springer Nature Switzerland AG
The registered company address is: Gewerbestrasse 11, 6330 Cham, Switzerland

To my uncle Mustafa İşlekçerçi, and my grandma 'Koca' Ayşe Özdemir who have both lost their lives during my Ph.D. studies.

Supervisor's Foreword

Graphene has long established itself as a playground to investigate topological effects in condensed matter physics: from the non-trivial Berry phase in mono- and bi-layer graphene to the flat bands with non-trivial Chern numbers in various twisted graphene heterostructures. This also prompts the observation of a number of many-body phenomena in such systems.

Dr. Servet Ozdemir's thesis expands the study of topological and many-body phenomena in graphene-based systems. He focused on the study of electronic properties in thin rhombohedral graphite devices, where electronic bands demonstrate very untrivial topology. Rhombohedral graphite is probably the simplest topological insulator with gapped bulk and conducting surfaces. Furthermore, the electronic states at the surface are sublattice polarised, so transverse electric field can open a gap in the electronic nature.

Because of the topological nature of the surface state, observation of the quantum Hall effect was possible in such devices. Furthermore, at low temperatures a spontaneous gap is opened in the surface state spectrum, which results in insulating state and pronounced hysteresis.

Servet's thesis offers insights to a number of long-standing questions on the nature of electronic states in rhombohedral graphite, having reported its first systematic electron transport investigation. More importantly, it opens an avenue for many further investigations into the nature of the correlated states observed and, because of many possible tuning parameters, offers opportunities in construction of novel correlated states in such systems.

This thesis is an example of dedicated and ingenious work of a talented physicist, supported by a team of very strong collaborators. No doubt we will see a continuation of the story of the rhombohedral graphite and of Servet himself.

Singapore
June 2021

K. S. Novoselov

Preface

This Ph.D. thesis, being about the first systematic study of electronic properties of mesoscopic rhombohedral graphite thin films encapsulated by hBN crystals, lies at the interface of graphene physics, van der Waals heterostructure devices and topological matter.

Typically, within the condensed matter physics group of the University of Manchester, each Ph.D. student specialising in electron transport experiments has a range of different samples to measure during their Ph.D., therefore being involved in many projects. Hence, the story of my Ph.D. thesis began with one of those samples, which was a 10-layer rhombohedral graphite film Hall bar. Despite being very new to the field, I was immediately able to realise that the project held a huge prospect, given the exotic predictions on the low energy bands of rhombohedral graphite, and the topological properties the system possesses. For three years, we were able to identify rhombohedral domains in exfoliated graphite films of up to 50-layer thickness, with the experiments having been resulted in very impactful publications demonstrating this system as an N-layer generalisation of graphene in low energy limit. Having being part of the team carried out this first systematic investigation into rhombohedral graphite films, I thought it would have been the most coherent story out of all the projects I have been involved in during my Ph.D, which are at various stages. Moreover, being at the centre of the experiments, and involved in heavily in the discussions of fabrications of the samples, as well as the theory calculations meant that I had become naturally one of the most qualified people to write a monogram on the topic, which is now happening with my Ph.D. thesis thanks to the nomination of my main supervisor Sir Konstantin Novoselov to the Springer thesis program.

Rhombohedral graphite thin films or ABC stacked multilayer graphene systems had been attracting a lot of theoretical interest due to their predicted extremely flat bands in the low energy limit which were in principle susceptible to various many-body interactions such as magnetism and superconductivity. However, due to the competing Bernal hexagonal phase being more energetically favourable, experimental investigations into rhombohedral stacked graphite films had been limited until our efforts on the investigation of this matter which is the subject of this thesis.

Chapter 1 is a general introduction of the thesis that reviews existing work on rhombohedral graphite systems that are both theoretical and experimental. The chapter introduces tight binding approach of modelling graphene and rhombohedral graphite, which is the most powerful method for modelling this system. Central concepts to graphene quantum transport such as pseudospin, valley, Berry phase and chirality are discussed which are shown to be also equally valid for rhombohedral graphite films. Topological properties of rhombohedral graphite films are discussed where it is highlighted that the system is a 3D generalisation of 1D topological insulator model Su–Schrieffer–Heeger chain. The Landau-level spectra, interaction-induced gapped phases, such as spontaneous quantum Hall states and superconductivity, as well as a displacement field-induced gap in the low energy bands are discussed.

Chapter 2 is an introduction of electron transport. The concepts of effective mass, mobility and conductivity are discussed. Hall effect is introduced, as well as its quantum, anomalous and quantum anomalous analogues. Lastly, weak (anti)localisation is discussed as a very central phenomenon to the field of quantum transport.

Chapter 3 is dedicated to experimental details including and starting from fabrication of the devices, discussion of measurement electronics and control of the parameters such as temperature and magnetic field in the laboratory.

Chapter 4 shows that there is an interesting interplay of bulk and surface conduction in rhombohedral graphite as a function of temperature where a transition into a metallic phase takes place due to surface-dominated transport at low temperatures from bulk-dominated transport at high temperatures.

Chapter 5 discusses Landau-level spectroscopy of low energy surface states and the implied low energy band structure with a Berry phase of $N\pi$.

Chapter 6 is dedicated to a spontaneous gap opening at the low energy flat bands of 9–12 layer rhombohedral graphite films with peculiar symmetry properties.

Chapter 7 is about displacement field-induced gap opening in pristine rhombohedral stacked systems and how it is a signature of the absence of stacking faults.

Chapter 8 is dedicated to bulk charges and how they display three-dimensional transport due to the k_z momenta they possess albeit belonging to a thin film.

Chapter 9 summarises the thesis.

Edinburgh, UK
May 2021

Servet Ozdemir

Acknowledgements

Back in 2015, while struggling over the 3rd-year solid-state physics module taught at the University of Warwick by Prof. Rudolf Roemer, I would have never thought I would be doing my Ph.D. as the student of the guy who introduced the 'wonder' crystal graphene to the world of solid-state physics. Hence, first and foremost, I would like to acknowledge my main supervisor Sir. Konstantin Novoselov for accepting me as one of his Ph.D. students at the University of Manchester. I guess not every Ph.D. student has the chance to work for somebody who is the first reference in most of the published papers in their field of study. I am very grateful to Kostya for the support, patience and understanding he has shown in the most difficult times of my Ph.D., when I was battling some health issues. I hope to contribute now to what has become the world of 'van der Waals heterostructures' throughout my career and maybe one day become one of those names that he has introduced to this field.

Great physicists have alike great colleagues working with them, and in my case, I was very fortunate to learn everything about quantum transport experiments by one of those people, Prof. Artem Mishchenko. I am also very grateful to Artem for first his time and patience that he has provided me throughout my Ph.D. journey. As importantly, the experiments reported in this thesis were carried out as part of a project that was initiated by Artem. I am very thankful to him for involving me in this beautiful project, which allowed me to make it to a journal like Nature already at the very beginning of my career. I hope to keep our relationship (what is now a friendship I hope) long-lasting.

Of course, I am also very grateful to Prof. Irina Grigorieva and Prof. Vladimir Falko for accepting me as one of their students to the Graphene NOWNANO Centre of Doctoral Training. Here, I was lucky enough to be part of a cohort full of incredible people, whom I have learned a lot from also. I hope to stay in touch with them as one of their alumnae as well as the rest of the cohort as one of their friends. They were an essential part of my incredible experience here in Manchester. I am also very grateful to my transport, fabrication and theory colleagues from condensed matter physics group whom I have also learned a lot from.

And of course, my beautiful family. I know that, 20 years ago, my dad came over to this country to be able to provide us a better future. I hope that I was able to make him proud by making the best use of the opportunities I was given. My mum had her right to be educated taken away from her at a very early age, and I hope that I was able to make her proud by working for and finishing my Ph.D. with arguably the top scientists in the country. I will never forget the support from my brother Mehmet, both when I was in a bad state mentally and the times when I needed some financial help. I hope I was able to be a good academic role model for my sisters, Ozlem and Melissa.

It is difficult to be able to list everybody on a single page, but along the way I was lucky enough to meet many very incredible friends. I am very grateful to every single one of them. Many of my great teachers who have played a significant role in my education journey are of course unmentioned too. Finally, I am also very grateful to my extended family, who are mainly in the town of Golbasi of south-eastern Turkey. I know that they are already very proud of the first upcoming Ph.D. in the entire family.

Contents

1 Review of Rhombohedral Graphite 1
1.1 Introduction 1
1.2 Crystal Structure and Hexagonal Brillouin Zone of Graphene 3
1.3 Tight Binding Approximation of Graphene 4
1.3.1 Matrix Elements 5
1.4 Low Energy Bands 7
1.5 Linear Dispersion Relation and Massless Dirac Fermions 8
1.6 Pseudospin, Chirality, and Berry Phase 9
1.7 Tight Binding Model of AB-Stacked Bilayer Graphene 10
1.8 Low Energy Band Structure of Bilayer Graphene 11
1.9 Two Band Hamiltonian, Chirality, and Berry Phase in Bilayer Graphene 12
1.10 Tight Binding Model of ABC Stacked N-layer Graphene 12
1.11 Topological Arguments for Low Energy Flat Surface Bands 14
1.12 Landau Spectrum of ABC-Stacked Graphene Layers 17
1.13 Spontaneous Symmetry Broken States 19
1.14 Displacement Field Induced Gap and Screening Effects in Rhombohedral Stacks of Graphene 22
1.15 Trigonal Warping and Berry Phase of Nπ in Rhombohedral Graphite 23
1.16 Flat Band Superconductivity in Rhombohedral Graphite 25
1.17 Stacking Faults on Rhombohedral Graphite Films 26
1.18 Identification of Rhombohedral Graphite 29
1.18.1 Raman Spectroscopy 29
1.18.2 Transmission Electron Microscopy 32
1.19 Density Functional Theory Predicted Magnetic Gap 33
1.20 Angular Resolved Photoemission Spectroscopy of Rhombohedral Graphite 34
1.21 Shear Stress and Doping Induced Stacking Order Control 35
1.22 Summary 35
References 36

2 Fundamentals of Electron Transport ... 41
2.1 Introduction ... 41
2.2 Effective Mass, Mobility, and Electrical Conductivity ... 41
2.3 Ohms Law Using Boltzmann Transport Equation ... 43
2.4 Quasi Fermi Level Separation and Diffusive Electron Transport Viewpoint ... 47
2.5 Hall Effect in a Two-Dimensional System in the Presence of Non-quantising Magnetic Field ... 49
2.6 Multi-band Transport ... 50
2.7 Hall Effect in a Two-Dimensional System in the Presence of Quantising Magnetic Field ... 51
2.8 Hall Effects in the Absence of Magnetic Field ... 54
2.8.1 Anomalous Hall Effect ... 55
2.8.2 Spin Hall Effect ... 58
2.8.3 Valley Hall Effect ... 60
2.9 Localization Effects in Disordered Mesoscopic Systems ... 61
2.9.1 Classical Picture of Diffusion and Localization ... 61
2.9.2 Weak Localization ... 62
2.9.3 Weak Anti-localization ... 66
References ... 67

3 Experimental Technicalities ... 71
3.1 Fabrication of van der Waals Heterostructures ... 71
3.2 Measurement Electronics ... 73
3.2.1 Phase Sensitive Detection and Lock in Amplifier ... 74
3.3 Measurement Geometry and Electrostatic Gating ... 76
3.4 Temperature Control ... 78
3.4.1 Sorption Pump-Controlled Helium-3 Cooling ... 78
3.4.2 Helium-3/Helium-4 Mixture Dilution Fridge ... 80
3.5 Magnetic Field ... 81
3.6 Summary ... 83
References ... 83

4 Bulk Versus Surface Conduction in Rhombohedral Graphite Films ... 85
4.1 Introduction ... 85
4.2 Temperature Dependence of Zero Gate Resistivity ... 85
4.3 Multi Carrier-Type Transport ... 87
4.4 Semiconductor—Metal Transition ... 87
4.5 Summary ... 88
References ... 89

5 Landau Level Spectroscopy of Rhombohedral Graphite Films ... 91
5.1 Introduction ... 91
5.2 Single Gated Landau Fan Maps ... 91
5.3 Implied Low Energy Band Structure ... 92

5.4 Discussion 94
5.5 Summary 95
References 97

6 Spontaneous Gap Opening in at Charge Neutrality Point of Rhombohedral Graphite Films 99
6.1 Introduction 99
6.2 Thermal Activation Gap Accompanied by Topological Currents 100
6.3 DC Characterisation of the Gapped Resistive State 101
6.4 Magnetic Field Dependence 102
6.5 Hysteretic Behaviour 104
6.6 Summary 105
References 106

7 Displacement Field Induced Band Gap Opening in Rhombohedral Graphite Films 109
7.1 Introduction 109
7.2 Displacement Field Induced Resistivity Increase 109
7.3 Absence of a Gap Opening for Graphite Films of Mixed Stacking 112
7.4 Non-local Signal Due to Displacement Field Induced Gap Opening 114
7.5 Summary 116
References 117

8 Three Dimensional Quantum Interference of Bulk Electrons 119
8.1 Introduction 119
8.2 Magnetic Field Rotation of a 9-Layer Film 120
8.3 Magnetic Field Rotation of a 14-Layer Film 121
8.4 Summary 125
References 125

9 Summary 127

Curriculum Vitae 129

Chapter 1
Review of Rhombohedral Graphite

1.1 Introduction

Shortly after the discovery of quantum Hall effect by von Klitzing et al. [1] in 1980, topological nature of it was demonstrated by Thouless et al. [2] in 1982. It was found that realisation of quantization of Hall conductivity at various integers is represented by the celebrated Thouless-Kohmato-Nightengale-den Nijs invariant, which has later become known a Chern number. Before the turn of the same decade, in 1988, Haldane [3] showed that on a honeycomb crystal, with broken time-reversal symmetry, a quantum Hall effect may arise even in the absence of magnetic field, thus predicting what is known as anomalous quantum Hall effect.

A few decades later, with the discovery of graphene by Novoselov et al. [4] in 2004, the theoretical toy model that Haldane and many others used was now available in practice. Moreover, with the demonstration of its peculiar relativistic quantum Hall effect, associated Berry phase and valley degree of freedom graphene in the coming years was able to lead to an explosion research in field of meso and nanoscopic physics both experimentally and theoretically [5–7].

One of these early theory works, by Kane and Mele [8], suggested that the very weak spin orbit interaction in graphene may lead to a very small gap, in the presence of time-reversal symmetry, thus leading to a new quantized Hall effect with associated quantized spin currents. Unlike integer and anomalous quantized Hall effects which are characterised by Chern numbers, this new effect was found to be described with namely a different invariant, which has been dubbed as the Z2 invariant [9].

The generalisation of Kane and Mele's work to three dimensions has led to prediction of a new class of material, which were named as topological insulators [10–12]. Topological insulators were predicted to be insulating in bulk with low energy two-dimensional surface states having close analogies to graphene with a linear dispersion relation, and Berry phase [13, 14]. Later on classes of topological materials were expanded into Weyl and Dirac semimetals which were found to exhibit linear dispersion relation and relativistic quantum phenomena in 3D [15].

S. Ozdemir, *Electronic Properties of Rhombohedral Graphite*, Springer Theses,
https://doi.org/10.1007/978-3-030-88307-2_1

Meanwhile, the family of two-dimensional systems was also growing. Novoselov et al. [16] showed shortly after graphene that their scotch tape method was applicable to other van der Waals crystals, namely hBN, MoS_2, $NbSe_2$ and so on. Out of these materials, hexagonal boron nitride (hBN), also possessing a honeycomb crystal structure with a similar lattice parameter to graphene, was found to be insulating, possessing a large band gap even in monolayer limit. In 2010 Dean et al. [17] demonstrated the use of hBN as a substrate due to its remarkably smooth surface for high quality electronics which has kick started developments in what came to be known as van der Waals heterostructures.

Utilizing hBN as a substrate, in particular by forming a superlattice by aligning monolayer hBN and (mono/bi)-layer graphene crystals a fractal phenomenon in the electronic structure in the presence of magnetic field, called Hofstadter butterfly was observed [18–20]. Moreover, placing of graphene on hBN crystals was found to be generating a band gap (when the two crystals are aligned) leading to a finite Berry curvature and detection of topological currents in form of valley Hall effect [21]. Due to the self-cleansing mechanism and atomically smooth surface of hBN, crystals of 2D materials are now widely encapsulated by hBN crystals, which has also enabled studies into air sensitive materials [22, 23].

Rhombohedral (ABC) stacked graphene systems have been subject to intense theoretical studies often being referred to as an N-layer generalisation of graphene, possessing flat surface bands with a Berry phase of $N\pi$ as well as $2N-2$ gapped bulk sub-bands which become 3D Dirac cones in bulk limit that spiral in k_z momentum space [24–31]. The flat bands have also been predicted to exhibit spontaneous quantum Hall states, as well as superconductivity [32, 33]. Rhombohedral stacks of graphene are an example of 3D generalisation of a Su–Schrieffer–Heeger chain, and belong to a new class of topological materials, namely the nodal line semimetals [34–36]. The flat bands arise as drumhead surface states whose boundaries are set by the projection of bulk nodal lines to the surface.

Despite these exotic predictions and the topological properties electronic transport investigation of rhombohedral graphite systems have been limited to tetralayers [37]. This thesis will explore first systematic investigation of electronic properties of rhombohedral graphite systems up to 50 layers thickness, in high quality, clean interface Hall bars of van der Waals heterostructures. In this introductory chapter we will summarise the existing theoretical and experimental work on rhombohedral graphite, using the simplest tight binding model of graphene as the starting point. We will adopt a structure very similar to the pedagogical introduction given by McCann in reference [38].

1.2 Crystal Structure and Hexagonal Brillouin Zone of Graphene

Graphene, the isolated single layer of graphite, is essential to understanding properties of its multilayers, possessing both hexagonal and rhombohedral phases [39, 40]. As a result, tight binding modelling of its band structure was carried out by Wallace [41], more than 70 years ago, long before its isolation in 2004.

Graphene has a honeycomb crystal structure and consists of a hexagonal Bravais lattice formed by A and B sub-lattices as shown in Fig. 1.1a, b. Primitive lattice vectors of the hexagonal Bravais lattice are

$$\mathbf{a}_1 = \left(\frac{a}{2}, \frac{\sqrt{3}a}{2}\right), \quad \mathbf{a}_2 = \left(\frac{a}{2}, \frac{-\sqrt{3}a}{2}\right) \tag{1.1}$$

as depicted in Fig. 1.1. The lattice constant, a, which is equal 2.46 Å is the distance between two neighbouring unit cells and the carbon–carbon bond length is equal to $a/\sqrt{3}$ (1.43 Å).

Reciprocal lattice vectors $\mathbf{b}_1$ and $\mathbf{b}_2$ related to the primitive lattice vectors by equation $\mathbf{a}_i \cdot \mathbf{b}_j = 2\pi\delta_{ij}$ $(i, j = 1, 2)$ are given by

$$\mathbf{b}_1 = \left(\frac{2\pi}{a}, \frac{2\pi}{\sqrt{3}a}\right), \quad \mathbf{b}_2 = \left(\frac{2\pi}{a}, \frac{-2\pi}{\sqrt{3}a}\right) \tag{1.2}$$

which are shown in Fig. 1.2 with the hexagonal Brillouin zone.

Each carbon atom making up graphene's honeycomb structure has 6 electrons. Out of these electrons 2 are core electrons and 4 of them are valance electrons occupying 2s, $2p_x$, $2p_y$ and $2p_z$ orbitals, respectively. In plane orbitals (2s $2p_x$ and $2p_y$) form the 3 σ bonds per carbon atom leading to a single electron ($2p_z$ orbital) per atom. This allows tight binding estimation to be made considering of one electron per atomic site.

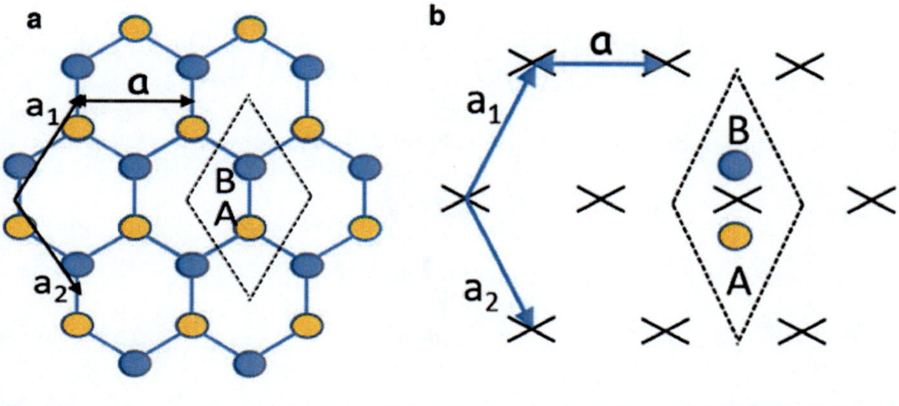

Fig. 1.1 **a** Hexagonal crystal structure of graphene consisting of A and B sublattices, **b** Bravais lattice of graphene. Vectors a_1 and a_2 are primitive lattice vectors and a is the lattice constant

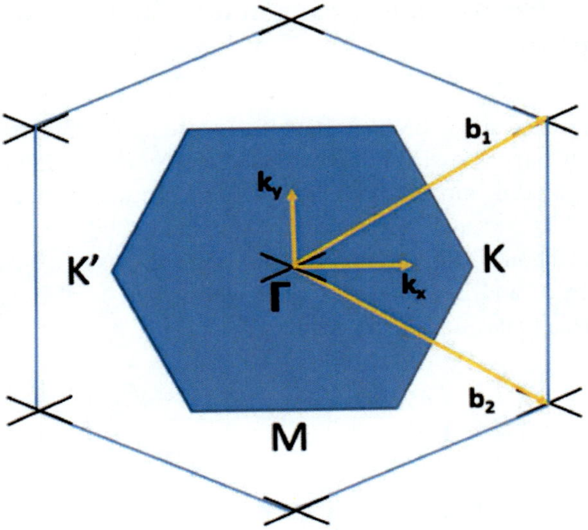

Fig. 1.2 Reciprocal lattice of graphene and the associated hexagonal Brillouin zone consisting of K and K′ valleys as well as M and Γ points. Vectors b_1 and b_2 are the reciprocal lattice vectors

1.3 Tight Binding Approximation of Graphene

An electronic eigenfunction describing a unit cell consisting of 2 single electron sites like the one of graphene is given as

$$\psi_j(\mathbf{k}, \mathbf{r}) = \sum_{i=1}^{2} C_{ji}(\mathbf{k}) \Phi_l(\mathbf{k}, \mathbf{r}) \tag{1.3}$$

where $C_{ji}(\boldsymbol{k})$ is a column vector and $\Phi_l(\boldsymbol{k}, \boldsymbol{r})$ is a Bloch function. Within tight binding method we can write a matrix equation as

$$\begin{pmatrix} H_{AA} & H_{AB} \\ H_{BA} & H_{BB} \end{pmatrix} \begin{pmatrix} C_{jA} \\ C_{jB} \end{pmatrix} = E \begin{pmatrix} S_{AA} & S_{AB} \\ S_{BA} & S_{BB} \end{pmatrix} \begin{pmatrix} C_{jA} \\ C_{jB} \end{pmatrix} \tag{1.4}$$

where the transfer integral matrix elements are defined as

$$H_{ij} = \langle \Phi_i \mathrm{H} \Phi_j \rangle \tag{1.5}$$

and the overlap integral matrix elements are defined as

$$S_{ij} = \langle \Phi_i \Phi_j \rangle \tag{1.6}$$

leading to the secular equation

$$\det\begin{pmatrix} H_{AA} - ES_{AA} & H_{AB} - ES_{AB} \\ H_{BA} - ES_{BA} & H_{BB} - ES_{AA} \end{pmatrix} = 0 \tag{1.7}$$

which one needs to solve to be able to obtain the dispersion relation.

1.3.1 *Matrix Elements*

To be able to solve the secular equation given in (1.7) matrix elements must be evaluated. Going back to the definition of an eigenfunction in (1.3), the Bloch function $\Phi_l(\boldsymbol{k}, \boldsymbol{r})$ is defined as

$$\Phi_l(\mathbf{k}, \mathbf{r}) = \frac{1}{\sqrt{N}} \sum_{i=1}^{N} e^{i\mathbf{k}\cdot\mathbf{R}_j} \phi_j(\mathbf{r} - \mathbf{R}_j) \tag{1.8}$$

where ϕ_j is jth atomic orbital in ith unit cell and the sum is over N unit cells.

1.3.1.1 Diagonal Elements

Following the definition above, the diagonal element H_{AA} would be evaluated as

$$H_{AA} = \frac{1}{N} \sum_{i=1}^{N} e^{i\mathbf{k}\cdot(\mathbf{R}_{A,i} - \mathbf{R}_{B,i})} \langle \phi_A(\mathbf{r} - \mathbf{R}_{A,i}) | \mathrm{H} | \phi_A(\mathbf{r} - \mathbf{R}_{A,i}) \rangle \tag{1.9}$$

given that there is only one A sublattice in every unit cell, Eq. (1.9) reduces to

$$H_{AA} = \frac{1}{N} \sum_{i=1}^{N} \langle \phi_A(\mathbf{r} - \mathbf{R}_{A,i}) | \mathrm{H} | \phi_A(\mathbf{r} - \mathbf{R}_{A,i}) \rangle = \frac{1}{N} \sum_{i=1}^{N} \varepsilon_{2p_z} = \varepsilon_{2p_z} \tag{1.10}$$

where ε_{2pz} is equal to the onsite energy of the p_z orbital in every A sublattice.

Similarly, diagonal element H_{BB} becomes, given the inversion symmetry of the system,

$$H_{BB} = \frac{1}{N} \sum_{i=1}^{N} \langle \phi_B(r - R_{B,i}) | \mathrm{H} | \phi_B(r - R_{B,i}) \rangle = \frac{1}{N} \sum_{i=1}^{N} \varepsilon_{2p_z} = \varepsilon_{2p_z} \tag{1.11}$$

where ε_{2pz} is equal to the onsite energy of the p_z orbital in every B sublattice.

Overlap of orbitals corresponding to diagonal elements S_{AA} and S_{BB} is equal to a unity,

$$\langle\phi_A(r - R_{A,i})|\phi_A(r - R_{A,i})\rangle = \langle\phi_B(r - R_{B,i})|\phi_B(r - R_{B,i})\rangle = 1 \tag{1.12}$$

given that the there is a single orbital on the same lattice site. This leads to the diagonal overlap elements being equal to

$$S_{AA} = S_{BB} = \frac{1}{N}\sum_{i=1}^{N} 1 = 1 \tag{1.13}$$

and hence to be equated to a unity.

1.3.1.2 Off-Diagonal Elements

Following from the definition of Bloch function in Eq. (1.8) and the transfer integral matrix in Eq. (1.5), the off-diagonal element describing electron hopping from A sublattice to surrounding three B sublattices is

$$H_{AB} \approx \frac{1}{N}\sum_{i=1}^{N}\sum_{l=1}^{3} e^{i\mathbf{k}\cdot(\mathbf{R}_{B,l}-\mathbf{R}_{A,i})}\langle\phi_A(\mathbf{r} - \mathbf{R}_{A,i})|H|\phi_B(\mathbf{r} - \mathbf{R}_{B,l})\rangle \tag{1.14}$$

where the term $< \phi_A|H|\phi_B >$ is equated to a positive quantity as $\gamma_0 = -\langle\phi_A|H|\phi_B\rangle$ which is known as the hopping parameter. This leads to Eq. (1.14) being written as

$$H_{AB} = -\frac{1}{N}\sum_{i=1}^{N}\sum_{l}^{3} e^{i\mathbf{k}\cdot(\mathbf{R}_{B,l}-\mathbf{R}_{A,i})} = \frac{-\gamma_0}{N}\sum_{i=1}^{N}\sum_{l}^{3} e^{i\mathbf{k}\cdot(\delta_l)} \approx -\gamma_0 f(\mathbf{k}) \tag{1.15}$$

where $f(\boldsymbol{k}) = \sum_{l=1}^{3} e^{i\boldsymbol{k}.\delta_l}$ with δ_l being equal to position vectors of three B sublattice atoms relative to A sublattice atoms.

The function $f(\boldsymbol{k})$ obtained using the position vectors δ_l simplifies to

$$f(\mathbf{k}) = e^{ik_y a/\sqrt{3}} + 2e^{-ik_y a/2\sqrt{3}}\cos(k_x a/2) \tag{1.16}$$

analogous to element AB, transfer integral matrix element BA is written as

$$H_{BA} = -\gamma_0 f^*(\mathbf{k}) \tag{1.17}$$

where $f^*(\boldsymbol{k})$ is a complex conjugate of the function $f(\boldsymbol{k})$.

Similarly, off diagonal integral element AB is calculated as

$$S_{AB} \approx \frac{1}{N}\sum_{i=1}^{N}\sum_{l=1}^{3} e^{i\mathbf{k}\cdot(\mathbf{R}_{B,l}-\mathbf{R}_{A,i})}\langle \phi_A(\mathbf{r}-\mathbf{R}_{A,i})|\phi_B(\mathbf{r}-\mathbf{R}_{B,l})\rangle = s_0 f(\mathbf{k}) \tag{1.18}$$

where s_0 is a non-zero parameter taking care of the fact that orbitals on neighbouring atomic sites may not be orthogonal. Analogous to integral element AB, integral element BA is written as

$$S_{AB} \approx s_0 f^*(k) \tag{1.19}$$

1.4 Low Energy Bands

Using the matrix elements calculated in the previous sections, solving the secular equation in (1.7) one obtains the energy equation as (equating ε_{2p_z} to 0)

$$E_{\pm} = \frac{\pm\gamma_0|f(\mathbf{k})|}{1 \mp s_0|f(\mathbf{k})|} \tag{1.20}$$

leading dispersion plotted in Fig. 1.3a with b showing a line cut through the band structure.

The values of the tight binding parameters γ_0 and s_0 do not affect the main qualitative features of the band structure plotted above. Here they have been taken to be 3.033 eV and 0.129, respectively. The values of the actual parameters for tight

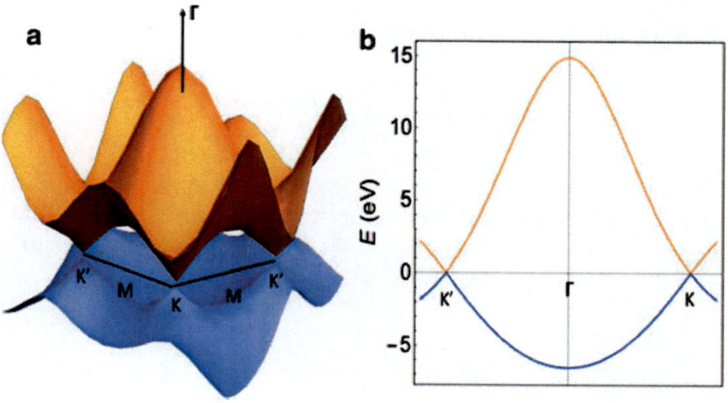

Fig. 1.3 **a** Three-dimensional band structure of graphene with K, K′ valleys and the Γ and M points depicted. **b** A two-dimensional cut of the graphene band structure from K and K′ valleys across the Brillouin zone

binding model of graphene and its multilayers have been attempted in various experiments [42–44]. For reference, they have been taken, in the calculations reported in this thesis as $\gamma_1 = 0.38$ eV, $\gamma_2 = -0.02$ eV, $\gamma_3 = 0.38$ eV, $\gamma_4 = 0.14$ eV.

1.5 Linear Dispersion Relation and Massless Dirac Fermions

As shown in Figs. 1.2 and 1.3a, six Brillouin zone corners of graphene are composed of 3 pairs of equivalent K and K′ points which are non-equivalent to each other. In the rest of this section, index, $\alpha = \pm 1$ will be used to distinguish between K and K′ valleys. Considering the reciprocal lattice vectors given in Eq. (1.2) and the respective Brillouin zone diagram in Fig. 1.2, the reciprocal lattice point for K (K′) point is

$$\mathrm{K}^{\alpha} = \pm\left(\frac{4\pi}{3a}, 0\right) \quad (1.21)$$

by substituting these lattice points to Eq. (1.16) one finds that the function $f(\boldsymbol{k})$ is cancelled to exactly zero, indicating absence of coupling between two sub-lattices at K and K′ points.

Introducing a momentum that is measured from the centre of K, $k_x + \frac{4\pi}{3a}$, one obtains the function $f(\boldsymbol{k})$ to be as

$$f(\mathbf{k}) = -\frac{\sqrt{3}a}{2\hbar}(p_x - ip_y) = -\frac{\sqrt{3}a}{2\hbar}\pi \quad (1.22)$$

and assuming onsite lattice energies ε_{2p_z} to be zero, using Eq. (1.22) one obtains the transfer integral matrix as in the vicinity of K point as

$$H = v\begin{pmatrix} 0 & p_x - ip_y \\ p_x + ip_y & 0 \end{pmatrix} \quad (1.23)$$

where the quantity ν is Fermi velocity of graphene equal to $\frac{\sqrt{3}a\gamma_0}{2\hbar}$. Assuming a unitary overlap integral matrix, the eigenvalues obtained from (1.23) lead to dispersion relation

$$\varepsilon(\mathbf{k}) = \pm v\sqrt{p_x^2 + p_y^2} = \pm v\mathbf{p} \quad (1.24)$$

with eigenvectors leading to the Bloch function for A and B sublattices respectively as

$$\psi = \frac{1}{\sqrt{2}}\begin{pmatrix} 1 \\ \pm \mathrm{e}^{\pm i\theta} \end{pmatrix} \mathrm{e}^{i\mathbf{p}\cdot\mathbf{r}/\hbar} \tag{1.25}$$

where $\boldsymbol{\theta}$ is the angle of momentum which is in x–y plane, $\boldsymbol{p} = \boldsymbol{p}(\mathbf{cos}\boldsymbol{\theta}, \mathbf{sin}\boldsymbol{\theta})$.

1.6 Pseudospin, Chirality, and Berry Phase

A remarkable property of the Bloch functions implied in (1.25) is that they resemble two component spinors describing a system consisting of either spin up or spin down states. However, unlike spinor functions, the eigenvectors leading to the Bloch function above arise because of the presence of A and B sub-lattices, in other words due to presence of 2 carbon atoms in the unit cell, thus making the honeycomb lattice structure of graphene consisting of 2 single electron sites, behaving as an analogue of spin and leading to the pseudospin degree of freedom. It can be seen that the pseudospins are also linked to momentum through the phase component, θ, which is a polar angle of the momentum in x–y plane thus leading the Bloch functions on the A and B sub-lattices being coupled to the direction of momentum and therefore resulting in chirality.

Since we have established the pseudospin freedom and its chirality, a convenient way of expressing the Hamiltonian in (1.23) at K and K′ valleys, is as

$$H = vp\big(\pm\sigma_x \cos(\theta) + \sigma_y \sin(\theta)\big) \tag{1.26}$$

where the Pauli spin matrices σ_x and σ_y are describing the pseudospin freedom arising from A and B sub-lattices.

Furthermore, evaluating the Bloch function in (1.25) for a closed circular loop, corresponding to a change in value of θ by 2π, a phase change that is equal to π is obtained as shown below

$$\pi = i \int_0^{2\pi} \mathrm{d}\theta\, \psi^* \frac{\mathrm{d}}{\mathrm{d}\theta} \psi \tag{1.27}$$

with the integral in Eq. (1.27) being known as the Berry phase integral [45] and the quantity of π obtained above being the experimentally verified characteristic Berry phase for graphene [5, 6].

1.7 Tight Binding Model of AB-Stacked Bilayer Graphene

An AB stacked bilayer system consists of 4 atoms in its unit cell as shown in Fig. 1.4a, which means it is described by 4 by 4 transfer and overlap integral matrices. The simplest tight binding framework is described by an additional hopping parameter to γ_0 which describes in plane hopping between nearest A and B lattice sites which is γ_1. Parameter γ_1 describes interlayer hopping between lattice sites A2 and B1 which lie directly on top of each other as shown in Fig. 1.4a. This hopping parameter is mathematically expressed by

$$\gamma_1 = \langle \phi_{A2}(\mathbf{r} - \mathbf{R}_{A,2})|\mathrm{H}|\phi_{B1}(\mathbf{r} - \mathbf{R}_{B,1})\rangle \tag{1.28}$$

thus, the transfer integral matrix which describe the lattice sites A1, B1, A2, B2 are given by

$$H = \begin{pmatrix} \varepsilon_{2p_z} & -\gamma_0 f(\mathbf{k}) & 0 & 0 \\ \gamma_0 f^*(\mathbf{k}) & \varepsilon_{2p_z} & \gamma_1 & 0 \\ 0 & \gamma_1 & \varepsilon_{2p_z} & -\gamma_0 f(\mathbf{k}) \\ 0 & 0 & \gamma_0 f^*(\mathbf{k}) & \varepsilon_{2p_z} \end{pmatrix} \tag{1.29}$$

and the overlap integral matrix is given as

$$S = \begin{pmatrix} 1 & -s_0 f(\mathbf{k}) & 0 & 0 \\ s_0 f^*(\mathbf{k}) & 1 & s_1 & 0 \\ 0 & s_1 & 1 & -s_0 f(\mathbf{k}) \\ 0 & 0 & s_0 f^*(\mathbf{k}) & 1 \end{pmatrix} \tag{1.30}$$

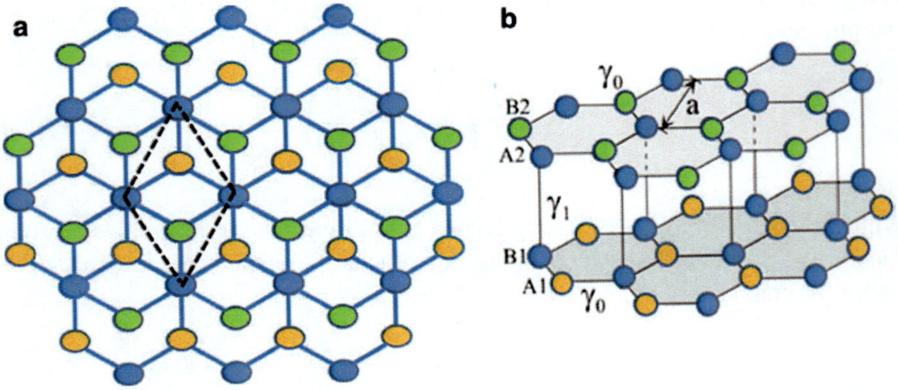

Fig. 1.4 **a** Hexagonal crystal structure of bilayer graphene consisting of four sublattices within the highlighted (dashed black lines) unit cell region. **b** Lattice sites A1, B1, A2, B2 within each unit cell illustrated as well as the hopping parameters γ_0 and γ_1

where s_1 is describes the non-orthogonality of orbitals on dimerised sites A1 and B2. Due to the small influence of parameters s_0 and s_1 on describing the low energy electronic properties they will be neglected, and a unitary overlap integral matrix will be assumed when discussing the low energy bands of AB-stacked bilayer graphene in Sect. 1.8.

1.8 Low Energy Band Structure of Bilayer Graphene

Assuming a unitary overlap matrix and solving the transfer integral matrix defined in (1.29) one obtains dispersion relation for 4 bands of a bilayer graphene system which is given by

$$E_{\pm}^{(\alpha)} = \pm\frac{\gamma_1}{2}\left(\sqrt{1+\frac{4\gamma_0^2 p^2}{\gamma_1^2}} + \alpha\right) \tag{1.31}$$

where $\alpha = \pm 1$. The dispersion relation, when $\alpha = 1$, implies two set of bands which are separated from each other by $2\gamma_1$ (depicted in orange on Fig. 1.5) as a result of dimerization of p_z orbitals in the vicinity of K/K′ points where $f(\mathbf{k})$ is equal to zero. When$\alpha = -1$, two set of touching bands in the vicinity of K and K′ points, which have a $E = -v^2 p^2/\gamma_1$ dispersion are obtained depicted in blue on Fig. 1.5. All bands, away from K and K′ points, have a linear dispersion in the large momentum limit.

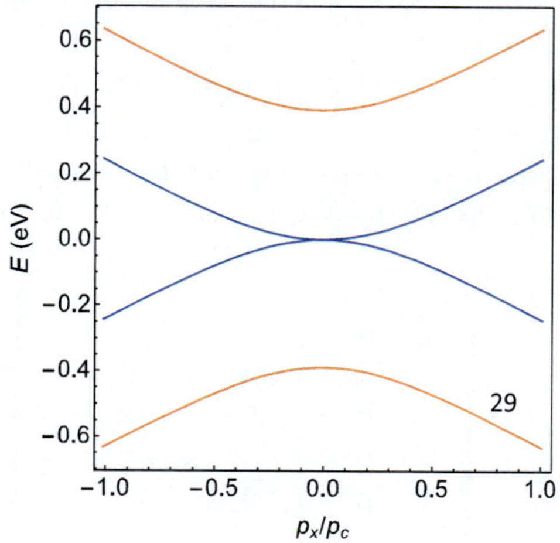

Fig. 1.5 Low energy band structure of bilayer graphene in the vicinity of K point. Orange bands are gapped because of dimerization of A and B sublattice orbitals that lie on top of each other. The lower energy bands arise from non-dimerised orbitals and they have a quadratic dispersion at low momentum

1.9 Two Band Hamiltonian, Chirality, and Berry Phase in Bilayer Graphene

Analogous to the case of single layer, the two band Hamiltonian of bilayer graphene describing massive chiral quasiparticles is given by

$$H = -\frac{v^2}{\gamma_1}\begin{pmatrix} 0 & (\pm p_x - ip_y)^2 \\ (\pm p_x + ip_y)^2 & 0 \end{pmatrix} \tag{1.32}$$

which gives us the low energy dispersion relation $E = \frac{v^2 p^2}{\gamma_1}$ with eigenfunctions leading to the Bloch function

$$\psi = \frac{1}{\sqrt{2}}\begin{pmatrix} 1 \\ \pm e^{\pm i2\theta} \end{pmatrix} e^{i\mathbf{p}\cdot\mathbf{r}/\hbar} \tag{1.33}$$

where the angle θ is the angle of momentum which is in x–y plane ($\boldsymbol{p} = p(\cos\theta, \sin\theta)$) as in the case of single layer graphene, which implies coupling of pseudospin degree of freedom to momentum as in the case of single layer graphene. However, inserting the above wavefunction into Berry phase integral would give a Berry phase of 2π instead of π for the case of monolayer graphene. The characteristic Berry phase of 2π in bilayer graphene has also been verified experimentally [46].

1.10 Tight Binding Model of ABC Stacked N-layer Graphene

The simplest Hamiltonian for an N-layer ABC stacked system considers of hopping parameters γ_0 and γ_1, analogous to bilayer graphene systems, however with a matrix dimension of $2N$ by $2N$ as shown in Eq. (1.34) in the vicinity of the K point.

$$H = \begin{pmatrix} 0 & v\pi & 0 & 0 & 0 & 0 & 0 \\ v\pi* & 0 & \gamma_1 & 0 & 0 & 0 & 0 \\ 0 & \gamma_1 & 0 & v\pi & 0 & 0 & 0 \\ 0 & 0 & v\pi* & 0 & \gamma_1 & 0 & 0 & \cdots \\ 0 & 0 & 0 & \gamma_1 & 0 & v\pi & 0 \\ 0 & 0 & 0 & 0 & v\pi* & 0 & \gamma_1 \\ & & & & \cdots & & \end{pmatrix} \tag{1.34}$$

where the function π and π*are defined identical to the single and bilayer systems.

Low energy bands (depicted in blue on Fig. 1.6a–c) of an N-layer ABC graphene system in the vicinity of K point is described by the two band Hamiltonian [27]

$$H = -\frac{v^N}{\gamma_1^{N-1}} \begin{pmatrix} 0 & (\pm p_x - i p_y)^N \\ (\pm p_x + i p_y)^N & 0 \end{pmatrix} \tag{1.35}$$

leading to the low energy dispersion relation

$$E = \pm \frac{(vp)^N}{\gamma_1^{N-1}} \tag{1.36}$$

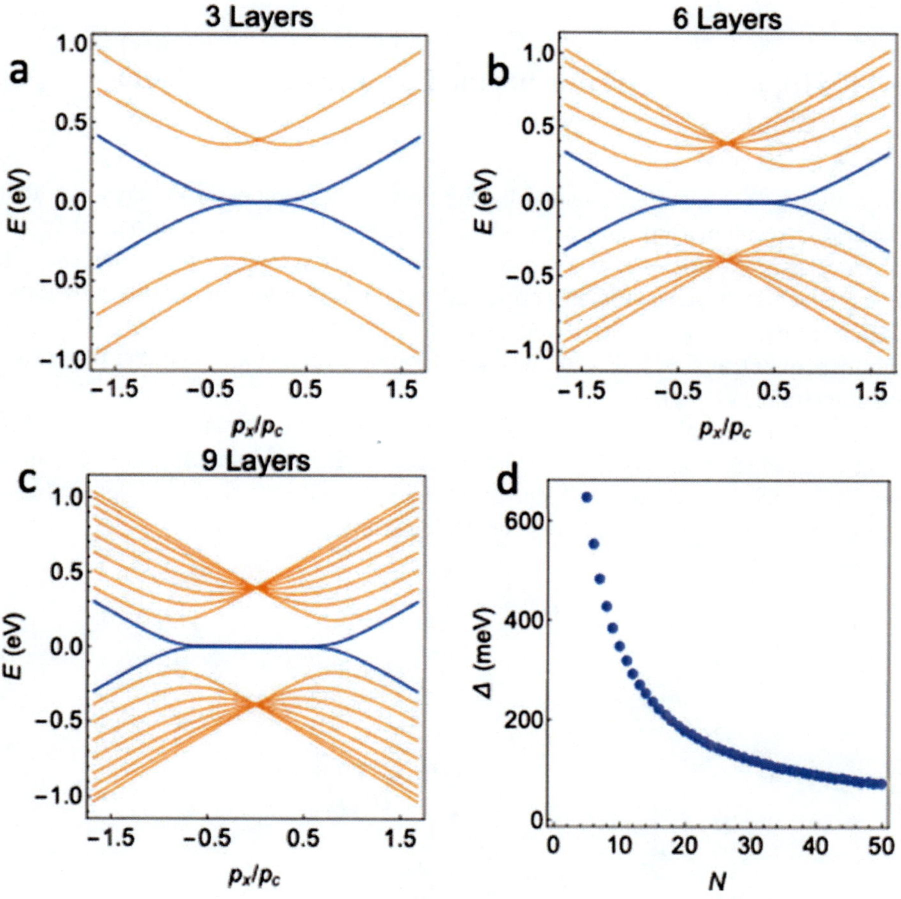

Fig. 1.6 **a–c** Band Structure of ABC stacked graphene systems of 3, 6 and 9 layers respectively. It can be seen the surface bands depicted in blue are becoming flatter with increasing thickness whereas the bulk band gap is becoming smaller. **d** Size of the bulk band gap for increasing number of layers, a finite bulk gap is present for even 50 layers thickness

where the effective Hamiltonian above and the low energy dispersion relation are indeed a generalisation of bilayer graphene leading to a Berry phase of $N\pi$.

Further to the 2 low energy surface bands, there are also $2N - 2$ low energy, gapped, linearly dispersing bulk sub-bands which are depicted in orange on Fig. 1.6a–c. These bulk sub-bands have been found to be described by the analytical expression

$$\Delta^N_{gap} = 4\gamma_1 \sin\left(\frac{3\pi}{2(2N+1)}\right) \tag{1.37}$$

where N is the number of layers [47]. This expression is plotted on Fig. 1.6d for various number of layers. Smaller but finite bulk gap is present even for 50 layers.

1.11 Topological Arguments for Low Energy Flat Surface Bands

Topological systems are protected by the presence of a non-zero energy gap that occurs as a result of a topological phase transition which unlike conventional quantum phase transitions does not accompany a spontaneous symmetry breaking [48, 49]. Systems which have a symmetry protected topological order are described by topological numbers (Berry phase and associated winding number in 1D and Chern numbers in 2D, and Z2 index for the case of topological insulators) rather than order parameters [2, 11, 50].

One of the simplest examples of topological phase transitions lie in a 1D Su–Schrieffer–Heeger (SSH) chain [51]. An ABC stacked multilayer system could be mapped onto a 1D, becoming identical to a SSH chain or it could be regarded as a 3D generalisation of the SSH chain as depicted in Fig. 1.7 [34].

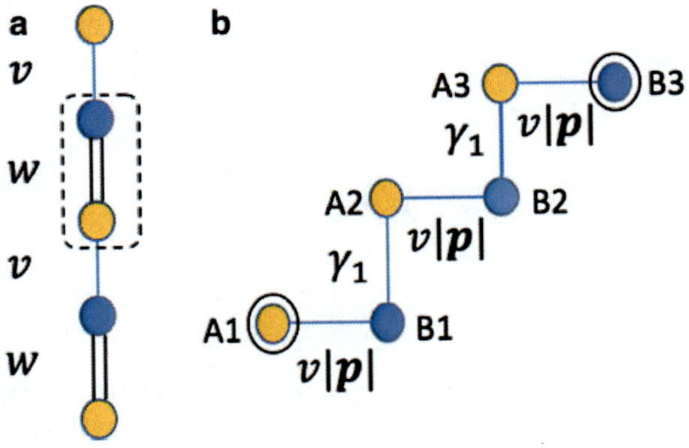

Fig. 1.7 **a** 1D Su–Schrieffer–Heeger chain of alternating hopping parameters w and v. **b** 3D ABC stacked graphene multilayers of in plane hopping parameter γ_0 leading to in plane Fermi velocity v and out of plane hopping parameter γ_1 leading to dimerised orbitals

The two band Hamiltonian for an ABC stacked system, once mapped onto a 1D chain becomes

$$H(\mathbf{k}) = \begin{pmatrix} 0 & v|\mathbf{p}| + \gamma_1 e^{i\mathbf{k}d} \\ v|\mathbf{p}| + \gamma_1 e^{-i\mathbf{k}d} & 0 \end{pmatrix} \tag{1.38}$$

with the Hamiltonian above yielding a dispersion relation

$$E = \pm\sqrt{(v|\mathbf{p}|)^2 + \gamma_1^2 + 2v|\mathbf{p}|\gamma_1 \cos(\mathbf{k}d)} \tag{1.39}$$

where it must be clarified that $|\mathbf{p}|$ is the magnitude in plane momentum within each graphene layer whereas $\mathbf{k}$ is the actual momentum along the chain with d being spacing of graphene layers. Eigenstates obtained through the dispersion relation are of the form

$$\psi(\mathbf{k}) = \frac{1}{\sqrt{2}} \begin{pmatrix} 1 \\ \pm e^{-i\phi(\mathbf{k})} \end{pmatrix} \tag{1.40}$$

where $\phi(\mathbf{k}) = \tan^{-1}\left(\frac{\gamma_1 \sin(\mathbf{k}d)}{v|\boldsymbol{p}| + \gamma_1 \cos(\mathbf{k}d)}\right)$. Following the eigenfunction defined above one can obtain the winding number through the Zak phase (Berry phase for a 1D system) as

$$\frac{1}{\pi} \int_{-\pi}^{\pi} i\psi * (\mathbf{k}) \frac{d}{d\mathbf{k}} \psi(\mathbf{k}) d\mathbf{k} = \begin{cases} 1 \; v|\mathbf{p}| < \gamma_1 \\ 0 \; v|\mathbf{p}| > \gamma_1 \\ \text{undefined } v|\mathbf{p}| = \gamma_1 \end{cases} \tag{1.41}$$

where the system is topologically non-trivial when the winding number is equal to 1. Another way of picturing the topological argument, which is demonstrated through Berry phase associated winding above, would be through introducing Pauli spin matrices to the Hamiltonian defined in (1.38). This leads to Eq. (1.42)

$$H(\mathbf{k}) = \begin{pmatrix} v|\mathbf{p}| + \gamma_1 \cos(\mathbf{k}d) \\ \gamma_1 \sin(\mathbf{k}d) \\ 0 \end{pmatrix} \begin{pmatrix} \sigma_x \\ \sigma_y \\ \sigma_z \end{pmatrix} = h(\mathbf{k}) \cdot \boldsymbol{\sigma} \tag{1.42}$$

where σ are the Pauli spin matrices and they are describing the sub-lattice degree of freedom rather than spin. Plotting the components $h(\mathbf{k})$ over the Brillouin zone for three separate cases as the momentum is tuned from π to $-\pi$ leads to the pictorial illustration of winding around the origin as shown in Fig. 1.8.

One can apply symmetry arguments to the Hamiltonian written in form above. The winding of the vector $h(\mathbf{k})$ around the origin is a consequence of the chiral symmetry (or otherwise known as sub-lattice symmetry). Mathematically one can define chiral symmetry as

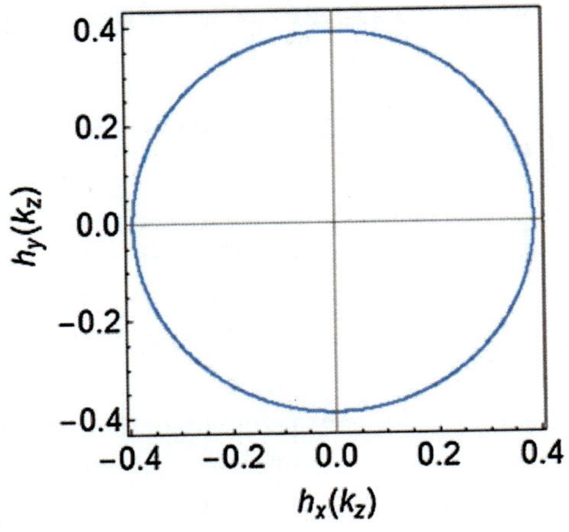

Fig. 1.8 Winding properties of vector $h(k)$ around the origin as $k_Z d$ is tuned from 0 to 2π. Above picture is broken when $v|p| \geq \gamma_1$

$$CHC^{-1} = -H \tag{1.43}$$

where C is a unitary matrix which anti-commutes with the Hamiltonian. Taking the Pauli spin matrix σ_z as the unitary operator one can satisfy the condition above as

$$\sigma_z H(\mathbf{k})\sigma_z^{-1} = -H(\mathbf{k}) \tag{1.44}$$

Using the above equation, one can immediately see that it is actually as a result of chiral symmetry that the vector $h(\mathbf{k})$ is restricted on the x–y plane and thus is able to have a defined winding number around the origin.

Moreover, considering wave-function amplitudes along the lattice sites one can gain insight into the edge or in the case of rhombohedral graphite films, surface property of the wave-functions. Figure 1.9 shows the wave-functions located at top and bottom layers for the case of low energy flat bands, whereas for the case of energies where the bulk sub-bands lie the wave-functions are found to be distributed within the gapped bulk.

Whilst sustaining chiral (sublattice) symmetry, in other words keeping the $h(\mathbf{k})_z$ component zero, the only possible way to change the winding number of the system from 1 to 0 is through closing the bulk gap and then reopening it which would be referred to as a topological phase transition, where the latter insulating phase would not be showing any surface states. It is at this transition point a really interesting example of bulk-boundary correspondence arises where a spiral nodal line in k_z direction could be mapped onto the 2D Brillouin zone to produce the flat bands of an ABC stacked N-layer system. One can re-write Eq. (1.39) for a 3D chain as

$$E^2 = (vp_x - \gamma_1 \cos(\mathbf{k}d))^2 + \left(vp_y - \gamma_1 \sin(\mathbf{k}d)\right)^2 \tag{1.45}$$

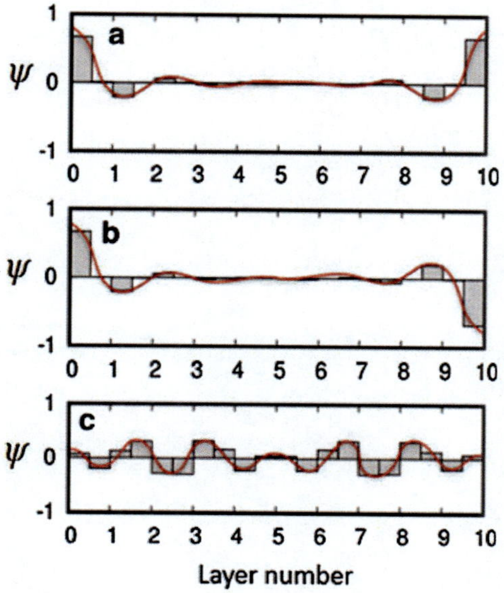

Fig. 1.9 Surface (**a**, **b**) and bulk (**c**) eigenstates of an ABC stacked chain, with wavefunctions located towards the edges for the two surface eigenstates

where zeros could be defined at

$$p_x = (\gamma_1/v)\cos(\boldsymbol{k}d) \text{ and } p_y = -(\gamma_1/v)\sin(\boldsymbol{k}d)$$

which leads formation of a spiral of zeros in 3D momentum state, which is referred to as a nodal line and is illustrated in Fig. 1.10 [35].

The mapping of the spiral in Fig. 1.10 leads to formation of a circle in the 2D p_x, p_y plane with a radius $|\mathbf{p}| = \gamma_1$ where the flat band exists. Thus, rhombohedral graphite in bulk form is a gapless a nodal line semimetal, which possesses flat bands in low energy limit that arise due to a projection of the Fermi surface to the 2D surface Brillouin zone [36]. It must be pointed out that in thin film limit, which are the systems studied in this thesis, rhombohedral graphite films are topological insulators.

1.12 Landau Spectrum of ABC-Stacked Graphene Layers

In the presence of a perpendicular magnetic field the Hamiltonian defined in Eq. (1.34) becomes modified as the in-plane momentum becomes perturbed due to the vector potential associated with magnetic field, $\mathbf{p} \rightarrow \mathbf{p} + \frac{e}{c}\mathbf{A}$. Rewriting the Hamiltonian defined in Eq. (1.34) around the K point using the raising and lowering operators $a = \frac{l(p_x - ip_y)}{\sqrt{2}\hbar}$ and $a^* = \frac{l(p_x + ip_y)}{\sqrt{2}\hbar}$ respectively, and

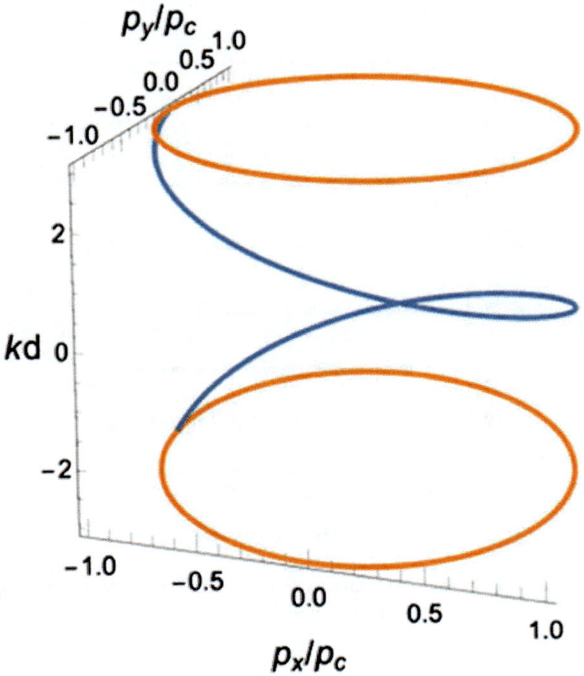

Fig. 1.10 Nodal line of a gapless rhombohedral stacked graphene system at the limit where the bulk band gap closes and its projection to the surfaces leading to flat bands with drumhead surface states

expanding the $2N$ component wavefunction defined on each sublattice of each layer, $(\psi_{A1}, \psi_{B1}, \psi_{A2}, \psi_{B2} \ldots, \psi_{AN}, \psi_{BN})$ as harmonic oscillator states $(\phi_0, \phi_1, \phi_2 \ldots)$, which are eigenfunctions of the operators a and a^* one is able to obtain the effective Hamiltonian in an nth Landau Level as [27]

$$H(n) = \begin{pmatrix} 0 & \varepsilon_n & 0 & 0 & 0 & 0 & \\ \varepsilon_n & 0 & \gamma_1 & 0 & 0 & 0 & \\ 0 & \gamma_1 & 0 & \varepsilon_{n+1} & 0 & 0 & \\ 0 & 0 & \varepsilon_{n+1} & 0 & \gamma_1 & 0 & \cdots \\ 0 & 0 & 0 & \gamma_1 & 0 & \varepsilon_{n+2} & \\ 0 & 0 & 0 & 0 & \varepsilon_{n+2} & 0 & \\ & & & \vdots & & & \end{pmatrix} \tag{1.46}$$

where $\varepsilon_n = \sqrt{2n}\hbar v/l$, with l being the magnetic length given by $l = 26\ \text{nm}/\sqrt{B}$.

Low energy Landau levels of an N-layer rhombohedral stacked graphene system has been obtained analytically to be [27, 29]

$$\varepsilon_n^{\pm} = \pm\hbar\omega_N\sqrt{n(n+1)\ldots(n+N-1)} \tag{1.47}$$

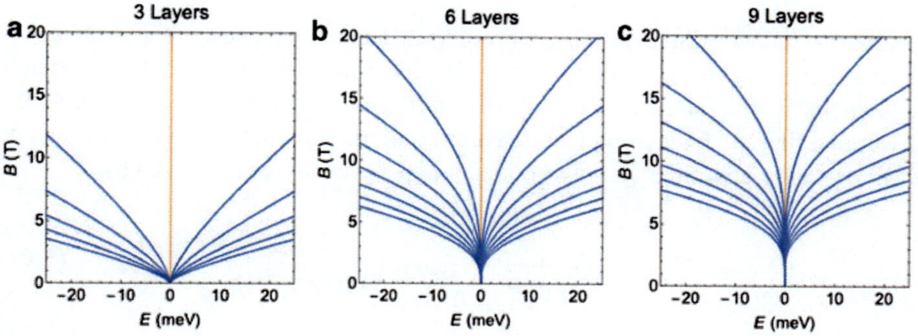

Fig. 1.11 Landau level spectrum of ABC stacked multilayer graphene flakes of **a** 3, **b** 6 and **c** 9 layers

for $n > 0$ where $\hbar\omega_N = \gamma_1(\sqrt{2}\hbar v/\gamma_1 l)^N$. It must be noted that there are N-zeroth Landau levels with the definition of a zeroth Landau level being the condition that the wavefunction that giving rise to the Landau level being present entirely on one surface. Figure 1.11 shows the low energy Landau levels for 3 layers, 6 layers and 9 layers ABC stacked systems taking only into account the hopping parameters γ_0 andγ_1. A clear change in magnetic field dependence of the Landau levels can be seen with respect to number of layers, as the magnetic field dependence of Landau levels with respect to number of layers, N, is $B^{N/2}$, pointed by Eq. (1.47). As well this, a finite onset magnetic field is found for the Landau levels as the layer number is increased [52]. At large energies beyond $\pm$ 25meVs, Landau levels start dispersing linearly with respect to the magnetic field [52, 53].

1.13 Spontaneous Symmetry Broken States

The low energy $E \sim p^N$ dispersion of rhombohedral graphene stacks suggested by the simplest tight binding considerations suggest the low energy bands to be increasingly flat as the number of layers increases. Given that symmetry broken states have already been demonstrated on bilayer systems, their dispersion property makes rhombohedral graphene stacks as ideal candidates for broken symmetry quantum phases [54–56]. The effective low energy gapped Hamiltonian describing the symmetry breaking for an N layer system has been found to be [32]

$$H_N = \frac{(vp)^N}{(-\gamma_1)^{N-1}}\left(\tau_v \cos(N\theta)\sigma_x + \sin(N\theta)\sigma_y\right) + \lambda\sigma_z \tag{1.48}$$

where $\cos(\theta)$ and $\sin(\theta)$ denote x and y components of momentum and $\tau_v = \pm 1$ denotes K or K′ valleys. Pauli matrices σ describe the pseudospin degree of freedom as in the case of single layer graphene, however also coupling to either one of the top or bottom layers in this case. The second term in the equation above describes the interaction induced gap between the low energy conduction and valance bands with 2λ being the magnitude of the band gap. The introduction of spontaneous inversion asymmetry by the interaction λ leads to a finite Berry curvature within the system. Assuming that the interaction is momentum independent, the present z-component of Berry curvature in the vicinity of conduction (valance) band is given by

$$\Omega_z = \pm \frac{\tau_v}{2} \frac{\lambda}{\left(\frac{(vp)^2}{\gamma_1^{N-1}} + \Delta^2\right)^{3/2}} \left(\frac{\partial}{\partial p} \frac{(vp)^{2N}}{\gamma_1^{2N-2}}\right) \tag{1.49}$$

where using the Berry curvature expression above, one can evaluate a corresponding orbital magnetisation and an intrinsic Hall conductance. The gapped phases are characterised by spin and valley dependence of the gap term $\lambda\sigma_z$ with regards to their orbital magnetisation and intrinsic Hall conductance. Overall, there are 4 ground states which are summarised in Fig. 1.12.

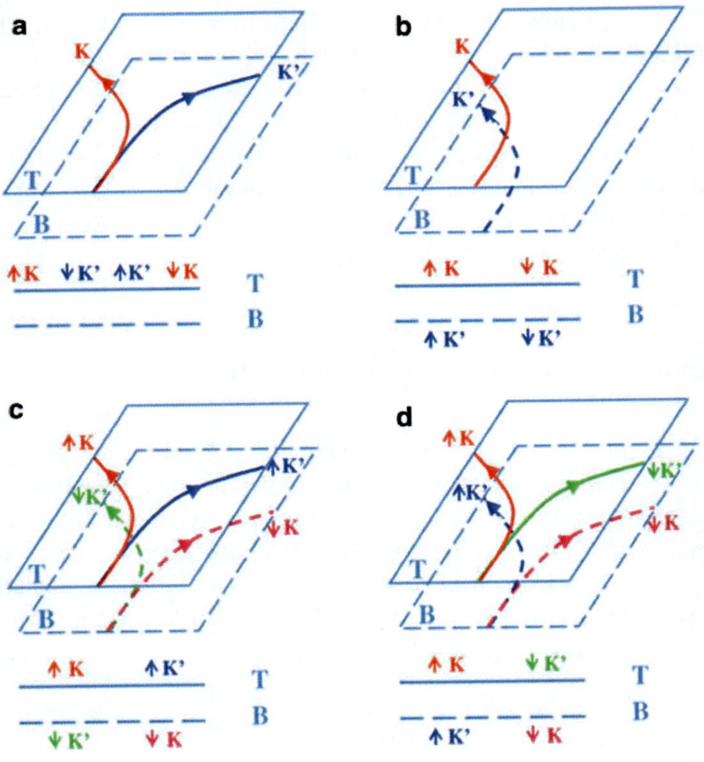

Fig. 1.12 **a** Layer polarised quantum valley Hall state and associated valley Hall currents, **b** Valley polarized quantum anomalous Hall insulator state with associated Hall currents, **c** Spin polarised layer antiferromagnetic state with associated spin polarized valley Hall currents at top and bottom surfaces d) Quantum spin Hall state with valley hall currents of opposite spins or spin currents of opposite valleys. Figure adapted from Zhang et al. [32]

The first of these states (Fig. 1.12a), the quantum valley Hall (QVH) insulator is a state where the interaction λ is of the form of a charge density wave, leading to a spontaneous net layer polarisation and opposite values of Berry curvature in two different valleys. The experimental observation of this phase would be through observation of a quantized valley Hall effect. Due to an overall cancelling of orbital magnetisation associated with Berry curvatures this phase preserves time reversal symmetry and has broken inversion symmetry only. The experimental observation of a quantum valley Hall phase has been hindered due to presence of inter-valley scattering events, which have destroyed the quantisation effects.

Quantum anomalous Hall (QAH) insulator (Fig. 1.12b) arises when the interaction λ is valley dependent, i.e. the interaction term in Eq. (1.48) is $\lambda\tau_v\sigma_z$ hence leading to a valley polarization. Due to the same Berry curvature and orbital magnetisation for each valley this state possesses an orbital magnetic moment and hence has broken time reversal symmetry. Its experimental observation would be as a quantised spontaneous Hall resistance present at even zero magnetic field. The Hall conductivity would be quantised with a layer dependent value of $2Ne^2/h$.

Layer antiferromagnetic insulator, illustrated in Fig. 1.12c, arises when the interaction λ is spin dependent, i.e. $\lambda s_z\sigma_z$, with $s_z = \pm 1$, for up and down spins orientation of spins. This spin dependent interaction leads to a spin polarised state, which has been demonstrated to be ground state in bilayer, and rhombohedral stacked multilayer experiments, utilizing suspended samples. This phase is expected to lead to spin filtered valley Hall currents, i.e. a valley Hall effect of opposite spins in top and bottom layers. This state also lacks time reversal symmetry due to its spin polarized, magnetic nature. The limitation brought by the suspended samples being two-probe has prohibited a detailed investigation of the layer antiferromagnetic phase.

The final potential ground state, topological insulator state arises when the interaction λ is dependent on both valleys and spins, i.e. when the interaction term becomes $\lambda s_z\tau_v\sigma_z$ and therefore leading to a valley polarisation with respect to same spins or spin polarization with respect to same valleys. If each layer is considered individually (which is the case in thick rhombohedral systems) this phase would lead to both spin and valley polarised currents in each layer. A clear distinction must be made with the quantum spin Hall phase specified here and the quantum spin Hall phase predicted and observed in TMDC systems or magnetic quantum wells. The quantum spin Hall state specified here has a broken Z2 symmetry and it arises with no consideration of spin–orbit coupling effects. Rather than a Z2 index its characterised by Chern numbers which arise from presence of finite Berry curvature. It is referred to as quantum spin Hall as the valley Hall currents are in opposite direction in opposite layers whereas the spin currents are in same direction in opposite layers. Therefore, in systems where top and bottom layers are electronically coupled, i.e. in thinner systems like bilayer, an overall quantum spin Hall Effect would be observed.

As well as the gapped symmetry broken states outlined above, there is also a possibility of gapless however rotation symmetry broken nematic state. Unlike the diagonal term λ in the case of gapped interaction, the presence of this phase is included through an off diagonal perturbation to the Hamiltonian [57]. This phase has been observed to be present in suspended bilayer systems and has also been identified as a possible symmetry broken state on rhombohedral stacked multilayer graphene systems. Its experimental manifestation is expected to show a drop in resistance at a filling factor $\nu = 0$ (in other words zero doping) at low magnetic fields.

1.14 Displacement Field Induced Gap and Screening Effects in Rhombohedral Stacks of Graphene

The low energy band structure of rhombohedral graphene stacks consists of flat bands which lie at the surface, as the bulk bands are gapped as a result of dimerization, leading to rhombohedral graphite stacks to behave as a zero gap semiconductor with diverging density of states. Aside from the interaction induced gapped phases, which could happen at zero doping/displacement field, rhombohedral graphite is also a platform which could exhibit single particle, displacement induced gap that is tuneable, as it has been observed on bilayer and tri-layer systems [58–65].

The effective low energy Hamiltonian in the presence of a potential difference between the first (U_1) and last (U_N) layers becomes

$$H = \begin{pmatrix} U_1 & \frac{v^N}{\gamma_1^{N-1}}(\pm p_x - i p_y)^N \\ \frac{v^N}{\gamma_1^{N-1}}(\pm p_x + i p_y)^N & U_N \end{pmatrix} \tag{1.50}$$

with $\Delta U = U_1 - U_N$ being the net potential difference applied across the sample, leading to the band gap.

Due to the diverging density of states, the flat bands can respond to any external electric field generated, by generating a carrier density imbalance, δn, which induces a screening field, $F_{screen} = \frac{-e\delta n}{2\varepsilon}$. Thus, the total potential difference across the system becomes equal to

$$\Delta U = e(F_0 + F_{screen})(N-1)d \tag{1.51}$$

where F_0 is the externally imposed electric field, N is the number of layers and d is the interlayer spacing of rhombohedral graphene stacks.

A self-consistent model taking into account of charge density imbalance, δn, has been found to lead to following approximate expression [58]

$$\Delta U \approx 2\gamma_1 F_0^{N/2}\left(\frac{e}{2\varepsilon}\frac{g_v g_s}{2\pi}\left(\frac{\gamma_1}{\hbar v}\right)^2 f_N\right)^{-N/2} \tag{1.52}$$

where

$$f_N = \frac{\Gamma(\frac{1}{2} - \frac{1}{N})\Gamma(1 + \frac{1}{N})}{2\sqrt{\pi}}$$

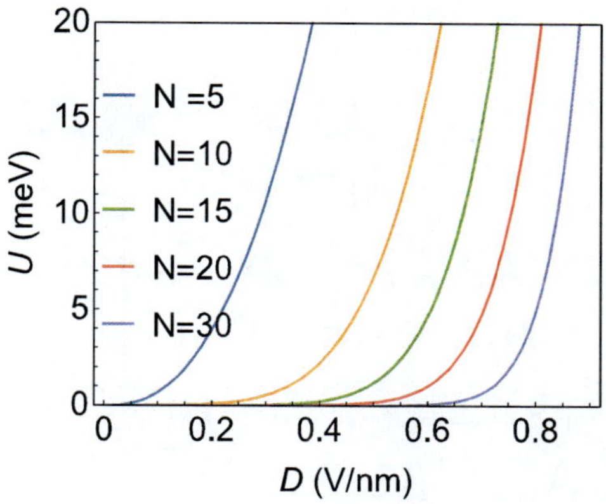

Fig. 1.13 Displacement field D induced gaps, U, on ABC stacked systems with thickness of 5, 10, 15, 20 and 30 layers respectively

with $\Gamma(x)$ being a gamma function. The f_N is equal to ½ in the large N limit which leads to the following simplified expression for displacement induced gap [58]

$$\Delta U \approx 2\gamma_1 \left(\frac{F_0}{F_C} \right)^{N/2} \tag{1.53}$$

where F_C is a critical field associated with a characteristic charge density of the flat band which is $\sim 1.2 \times 10^{13}$ cm^{-2}.

Figure 1.13 shows displacement induced gaps suggested by Eq. (1.53) for various number of layers of ABC stacked flakes. Greater displacement field is necessary for thicker number of layers to induce a sizeable gap detectable in electronic transport experiments.

1.15 Trigonal Warping and Berry Phase of $N\pi$ in Rhombohedral Graphite

The remarkably simplistic low energy dispersion scenario of $E \sim p^N$ (with N being the number of layers) is not the case when additional hopping parameters γ_2, γ_3 and γ_4 are considered which are depicted on Fig. 1.14a.

Once additional hopping parameters are considered, the flat bands of rhombohedral graphite are not as flat with charge carriers gaining a finite effective mass and the circular fermi surface undergoing a trigonal warping. Terms γ_2 and γ_3 are responsible for the trigonal distortion of the Fermi surface [29]. The term γ_4 breaks the electron–hole symmetry (or sometimes referred to as particle-hole symmetry), thus leading to the lifting of the topological protection described in Sect. 1.11 [66] albeit the low energy bands are still topological origin in a sense that they arise in the bulk gap due to topological properties of the gapped bulk bands. It is γ_4 which also leads to a quadratic low energy dispersion of the conduction and valance bands, leading to an effective mass.

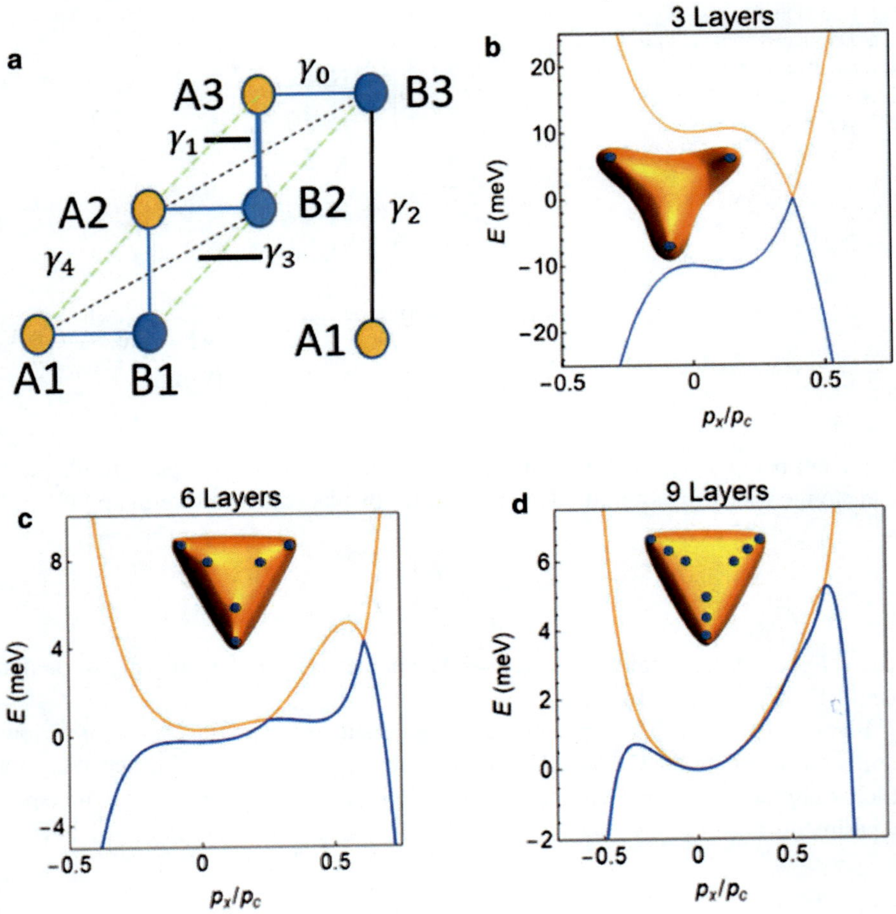

Fig. 1.14 **a** Tight binding hopping parameters γ_0 to γ_4 illustrated on an ABC stacked trilayer system. **b** Low energy band structure of ABC trilayer system showing one of the 3 band crossings illustrated in the inset. Trigonal warping of the Fermi surface can also be inferred from the inset. **c**, **d** Low energy band structures of 6- and 9-layer ABC stacked systems insets showing all band touching points and associated Berry phase of $N\pi$ as well as trigonal warping

The effective dispersion relation obtained by Koshino and McCann for an N-layer rhombohedral stacked system is

$$E = \frac{2vv_4 p^2}{\gamma_1} \pm |X(p)| \tag{1.54}$$

where

$$X(p) = \sum_{\{n_1,n_2,n_3\}} \frac{(n_1+n_2+n_3)!}{n_1!n_2!n_3!} \frac{1}{(-\gamma_1)^{n_1+n_2+n_3-1}} (vpe^{\pm i\phi})^{n_1} (v_3pe^{\mp i\phi})^{n_2} \left(\frac{\gamma_2}{2}\right)^{n_3}$$

and the summation is taken over $n_i \geq 0$ where $n_1 + 2n_2 + 3n_3 = N$. Fermi velocities v, v_3 and v_3 correspond to $\frac{\sqrt{3}a\gamma_0}{\hbar}$, $\frac{\sqrt{3}a\gamma_3}{\hbar}$ and $\frac{\sqrt{3}a\gamma_4}{\hbar}$ respectively. The effective mass obtained as a result of the quadratic dispersion around each K (K′) point at low momentum is $m_* = \frac{\gamma_1}{4vv_4} = 0.4m_e$.

Once the dispersion relation is plotted for the arbitrary number of layers, an interesting property is obtained with regards to the conduction and valance band crossings. The number of crossings which occur at K and K′ valleys at $X(p) = 0$ is an integer with respect to the number of layers. When $N + 1$ is not a multiple of 3, there are N linear conduction and valance band crossings with a Berry phase of π leading to a Berry phase of $N\pi$ even in the presence of trigonal warping. When $N + 1$ is a multiple of 3, there $N + 1$ Dirac points with a Berry phase of π and a Dirac point with a Berry phase of $-\pi$ also leading to an overall Berry phase of $N\pi$. The trigonal band distortion with respect to number of layers as well as the respective Dirac points is depicted in Fig. 1.14 for various numbers of layers. The presence of odd/even Berry phase with respect to odd/even number of layers could manifest itself in weak (anti)localization phenomena at quantum transport experiments in disordered limit [30].

1.16 Flat Band Superconductivity in Rhombohedral Graphite

The microscopic theory of superconductivity proposed by Bardeen, Cooper and Schrieffer (BCS) in 1957 was very successful in explaining many aspects of superconductors including the superconducting gap and its size, as well as the superconducting transition temperature. The BCS theory was successful in explaining superconductivity through a phonon mediated attractive interaction between electrons, W, which form pairs above the Fermi surface of a metal, lowering the ground state energy of the system and therefore leading to a gap with respect to the Fermi level of the metal as [67]

$$|\Delta| = 2E_C \exp\left(\frac{-1}{W\rho_F}\right) \tag{1.55}$$

where ρ_F is density of states at the Fermi level and E_c is the characteristic phonon energy. The negative exponential dependence on the interaction strength between electrons explains the usual low temperature transition temperatures for the onset of the superconducting state. From BCS theory one obtains the relationship between the superconducting energy gap and the critical temperature as [67]

$$2\Delta = 3.5T_C \tag{1.56}$$

where T_C is the transition temperature.

Unlike the expressions above, the flat bands offer a potential route to high temperature superconductivity through a linear dependence on characteristic pairing energy and the flat band area. Superconducting gap size of an ABC stacked system (although generalizable to other flat band systems) has been obtained to be as [33]

$$\Delta_0 = \frac{g}{8\pi} = \frac{Vp_{FB}^2}{\hbar^2} = 3k_B T_C \tag{1.57}$$

where p_{FB} is the momentum which defined the boundary of the flat band at the surface. Therefore, as well as being directly proportional to the interaction strength, the superconducting gap size and the transition temperature is directly dependent on the flat band area which suggests possible high temperature superconductivity.

Flat band superconductivity has recently been discovered on twisted bilayer graphene systems. Formed by twisting of graphene layers to 'magic angles' the flat bands have created a lot of excitement in solid state physics community. The superconductivity observed in 'magic angle' graphene bilayers has been dubbed to be unconventional however the transition temperatures have been found to be far from high temperatures with an initial value of 1.7 K, later on, with reported to increase to 3 K through application of pressure and following that with major improvements in the sample quality [68–70].

Superconductivity in rhombohedral graphite has also been predicted, when additional hopping parameters to γ_0 and γ_1 are considered and the topological protection of the flat bands is lifted, leading to a quadratic dispersion with a heavy effective mass. In this case an interesting interplay between flat band and BCS-like superconductivities has been proposed [66]. At low coupling energy, $g \ll (8\pi\gamma_4\gamma_1)/\gamma_0$, as well as assuming there is negligible doping, superconducting is found to be as

$$\Delta \approx \left(\frac{2\gamma_1\gamma_4}{\gamma_0}\right) \exp\left(\frac{-8\pi\gamma_1\gamma_4}{g\gamma_0}\right) \tag{1.58}$$

thus, showing an exponential dependence to the coupling strength, g, therefore leading to a BCS like superconductivity.

1.17 Stacking Faults on Rhombohedral Graphite Films

Given the hexagonal orientation of graphite being more thermodynamically stable, one is increasingly more likely to have stacking faults in rhombohedral stacks of greater number of layers. As a result, there have been a reasonable number of works, considering what happens to the electronic structure of graphene stacks in the presence of stacking faults [25, 71–75]. The number of different stacking possibilities have been estimated to be 2^{N-2} where N is the number of layers [26]. This means that when systems of large number of layers are considered simplifications had to

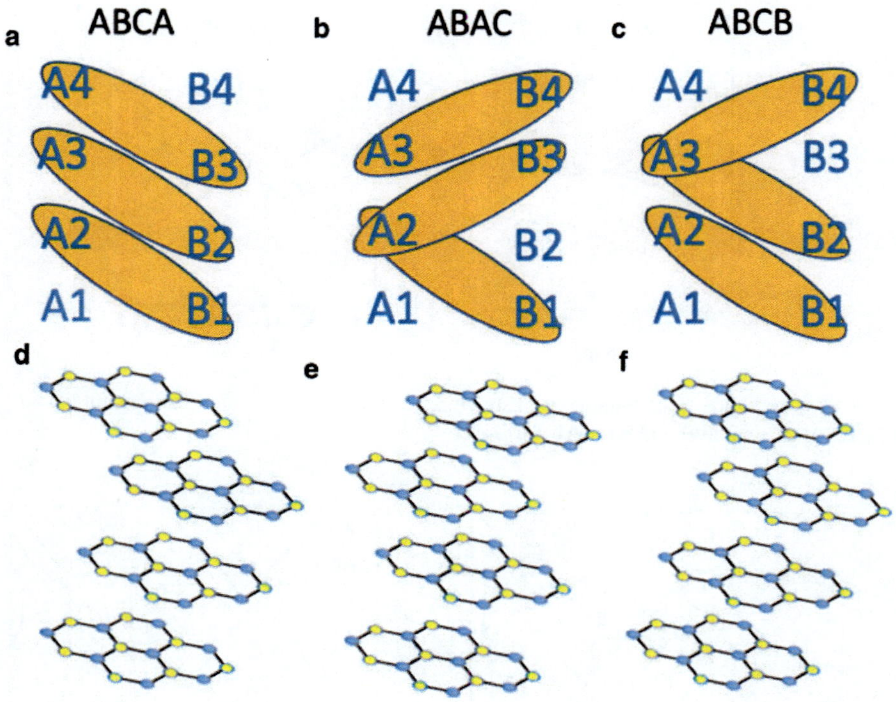

Fig. 1.15 Linkage diagrams and crystal structure schematics for **a**, **d** ABCA, **b**, **e** ABAC, **c**, **f** ABCB stackings

be made whereas for systems of smaller number of layers all possibilities have been studied.

Figure 1.15 shows orbital linkage diagrams and crystal stacking sequences for ABCA stacked tetra-layer system and its possible versions with stacking faults namely ABAC and ABCB. Linkage diagrams for ABAC and ABCB systems suggest the band structures in the vicinity of the K point to be identical which was indeed found to be case in tight binding calculations shown in Fig. 1.16 which are in agreement with the tight binding calculations carried out by Min and MacDonald [26, 73]. The band structures of ABAC and ABCB systems are very similar to one for ABA stacked tri-layer, possessing a superposition of 2 pairs of bands at zero energy, which are quadratically dispersing and linearly dispersing. The only difference to ABA stacking was found to be an additional pair of gapped sub-bands.

Prior to the tight binding calculations, two density functional theory (DFT) calculations were carried out for tetralayer systems possessing stacking faults. The first work was that of Latil and Henrad which predicted a band gap for ABAC stacking which is not in agreement with the tight binding calculations [71]. Second work by Aoki and Amawashi [72] considered tetralayer systems with stacking faults both in zero and finite displacement. Their calculations (shown in Fig. 1.17) in zero displacement field were in agreement with tight binding calculations and their prediction for

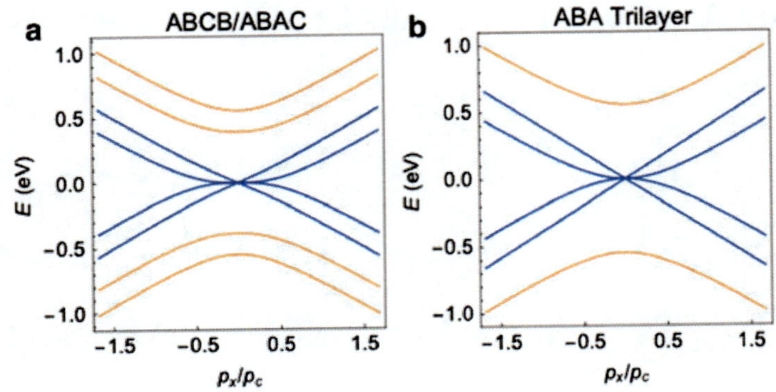

Fig. 1.16 Low energy tight binding band structures calculated in the vicinity of K point for **a** for ABCB/ABAC tetralayer and **b** for ABA tri-layer

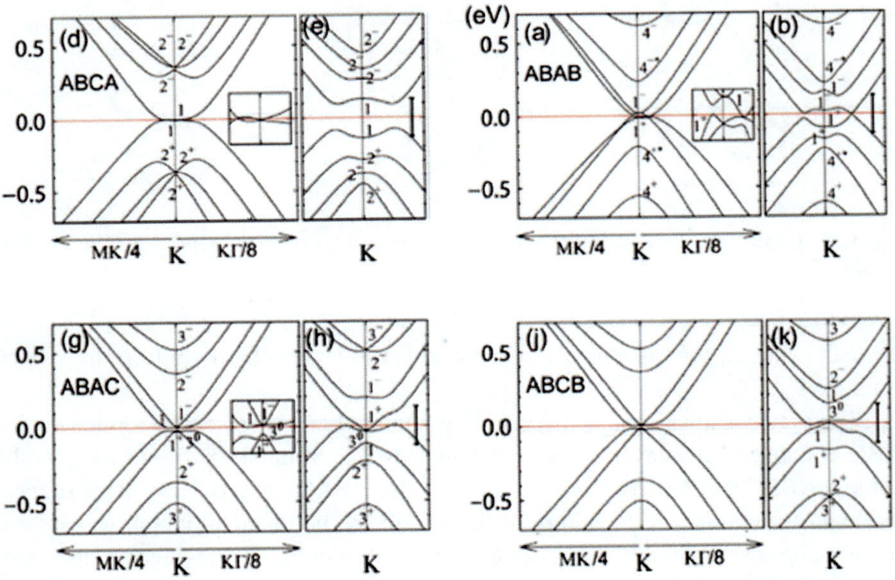

Fig. 1.17 Low energy band structures of tetra-layer stacking orientations at presence of zero and finite displacement field of 0.7794 V/nm for **a**, **b** ABCA, **c**, **d** ABAB, **e**, **f** ABAC, **g**, **h** ABCB structures respectively calculated using density functional theory Figure adapted from Aoki and Amawashi [72]

ABA stacked trilayer not exhibiting a band gap in the presence of finite displacement field was in agreement with experiments [65, 76]. The most striking prediction from this work is that tetralayer samples showing stacking faults were found not to open a band gap despite a large displacement field of 0.7794 V/nm. This prediction is yet to be tested in experiments and could in principle allow a potential method for verification of presence of stacking faults on rhombohedral stacked systems.

For systems of larger number of layers, an initial work utilizing tight binding calculations were carried out by Arovas and Guniea [25] where they considered a single stacking fault between two regions of hexagonal stacking, namely in form of ABABCBCB. They find that such kind of stacking faults separating two Bernal stacked regions leads to an additional low energy band with a cubic dispersion and would manifest itself in Landau fan spectroscopy experiments with Landau levels dispersing as $B^{3/2}$. Further to this study, a more systematic approach to study stacking faults was adopted by Koshino and McCann [74] where they considered several hexagonal stacked sections connected by regions of rhombohedral stacking. It is found that the overall systems are well approximated by the electronic structure of incomplete hexagonal stacked graphite regions which make up the overall system with rhombohedral stacking faults. The overall low energy spectrum was found to be comprised of linear, quadratic, and cubic bands.

Further to the tight binding calculations, a DFT work was carried out by Taut, Koepernik and Richter [75] going beyond the initial work on tetra-layer systems and considering two types of singular stacking faults on rhombohedral graphite slabs of structures $(ABC)_4(BCA)_4$ and $C(ABC)_4(BAC)_4$ as well as a surface stacking fault. A rhombohedral structure without stacking faults consists of dimers because of overlapping p_z orbitals. Out of the stacking faults considered, the former type was found to lead to two monomers and a tetramer, whereas stacking fault of the latter type was found to induce one monomer and a tetramer. Overall, it was found that the stacking faults lead to additional surface and interface low energy bands including linearly dispersing ones which form Dirac cones which has prevented a gap opening in the case of odd-numbered ABA stacked multilayer graphene.

Lastly, a very recent work carried out by Nery et al. [77] considers Bernal and rhombohedral stacking in graphite films, calculating energy distributions for each case as well as for multilayer systems with stacking faults. Although the calculations strikingly predicting rhombohedral phase becoming more energetically stable in comparison to Bernal phase, for very thick systems, it is concluded that both phases of Bernal, rhombohedral as well as mixed phases are expected to exist at room temperature as multilayer systems as it has been seen experimentally which has been the trigger point for this thesis.

1.18 Identification of Rhombohedral Graphite

1.18.1 Raman Spectroscopy

Identification of the low abundance (~15%) rhombohedral stacking on a few layer graphene flakes has been demonstrated to be possible through Raman scattering as well as far and near field optical microscopy [43, 78–80]. However, these works were limited to 6 rhombohedral graphite layers [78]. Recently, in a seminal work, Henni et al. [52] identified ABC stacked rhombohedral graphene flakes up to 17 layers

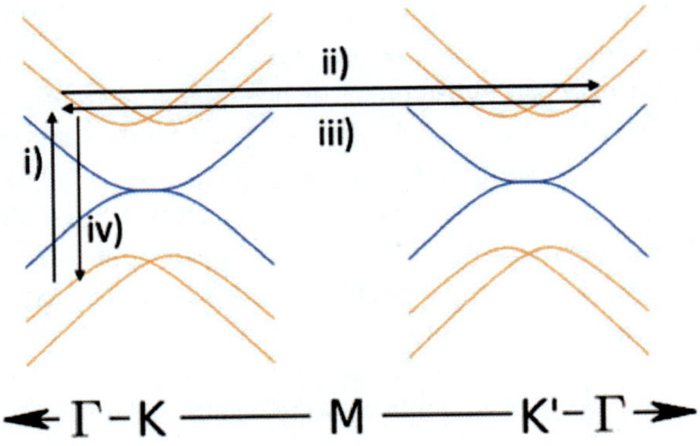

Fig. 1.18 Formation of the double resonant 2D peak on an ABC tri-layer system. (i) Creation of an electron hole pair, (ii) scattering of the electron to the K′ valley, (iii) Back scattering of the electron to the K valley, (iv) Recombination of the electron-hole pair

through Raman spectroscopy features including G-peak and particularly the double resonance 2D peak and verified the stacking through magneto-Raman scattering experiments.

The G peak arises from the first order process of in plane lattice distortions in graphene originating from degenerate in plane transverse optical and in plane longitudinal optical modes [81]. Double resonance 2D peak is however formed because of a 4-step process where two in plane transverse optical phonons with wavevector $\mathbf{k} \neq 0$ are involved. An electron–hole pair is initially created as a photon possessing energy of $\hbar\omega_L$ is absorbed. The electron (hole) is scattered with a momentum wavevector **k**. Following this, the electron (hole) is scattered back with a momentum wavevector -**k**. Final step is the recombination of the electron–hole pair [82]. The inelastic scattering of the electrons (holes) results in a link to the electronic structure of the system as shown in Fig. 1.18.

The summary of the Raman characteristics obtained for ~ 17 graphite layers by Henni et al. [52] is shown in Fig. 1.19. The two main features distinguishing ABC stacked domains from ABA ones is, broadening of the 2D peak, and a blue shift of the G-peak. As well as these two features a low intensity feature attributed to electronic Raman scattering across the bulk band gap appears between the G and 2D peaks as shown in Fig. 1.19e.

Following on from the work of Henni et al. [52], Torche et al. [83] carried out first principles DFT analysis for the Raman spectra of both rhombohedral and hexagonal stacked multilayer systems (Fig. 1.20a) as well as a system with stacking faults (Fig. 1.20b). The considered system with stacking faults was a periodic arrangement of rhombohedral and hexagonal tri-layers in form of [ABC](BAB)[CBA](BCB). The work suggested a definitive signature of the absence of the stacking faults which was found to be an additional kink at a laser energy dependent wavenumber located around the left end of the 2D peak as shown in Fig. 1.20a, b (at a wavenumber of ~ 2575 cm^{-1}).

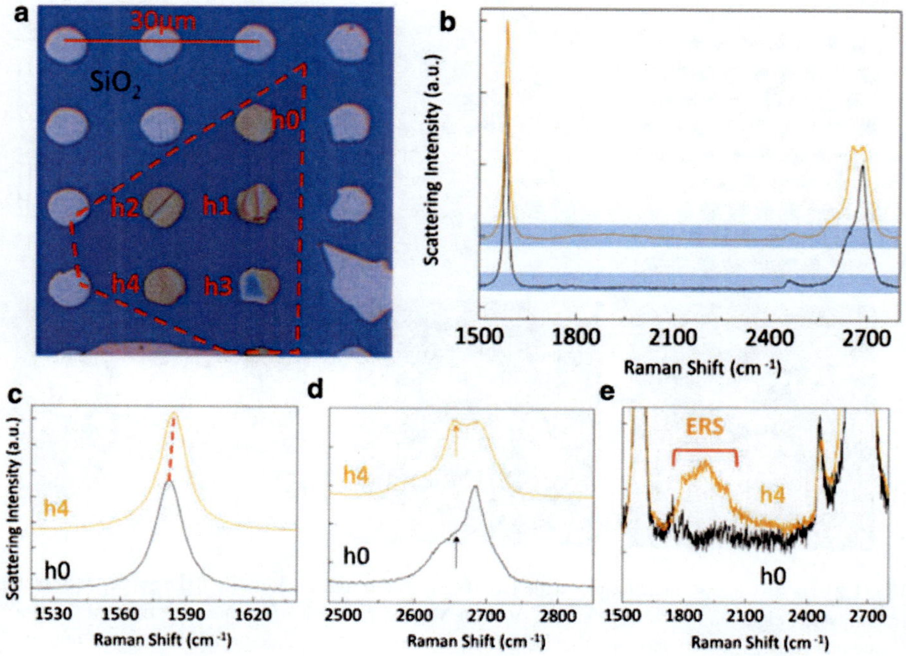

Fig. 1.19 **a** The studied ~ 17-layer system suspended over circular holes of 6 μm on a silicon substrate. **b** Raman spectra of the regions h4 (orange) and h0(black) containing ABC and ABA domains, respectively. **c** G peak of the Raman spectra shown in **b**. 2D peak of the Raman spectra shown in **b**. **e** Low intensity electronic Raman scattering feature and its absence on ABA domain. Figure adapted from Henni et al. [52]

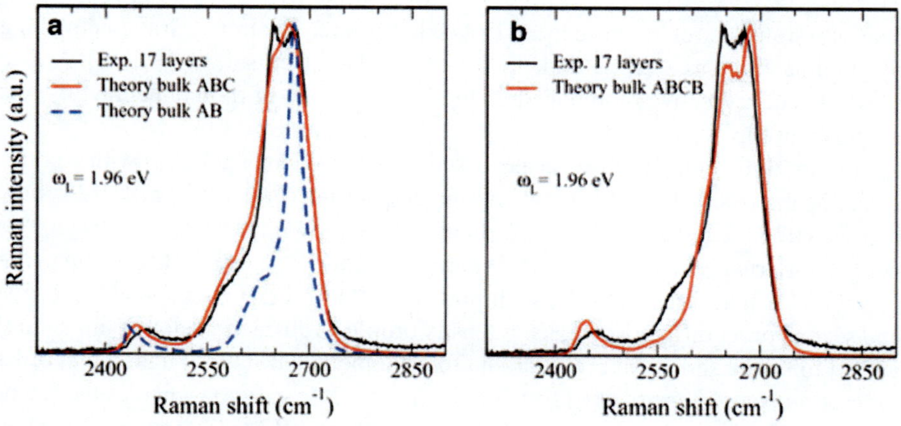

Fig. 1.20 **a** Experimentally obtained 2D peak for a 17 layer rhombohedral stacked system (black) as well as theoretical curves for bulk rhombohedral (red) and bulk hexagonal (dashed blue) stackings. **b** Experimentally obtained 2D peak for the 17-layer system (black) compared with 2D peak of a system with stacking faults. Figure adapted from Torche et al. [83]

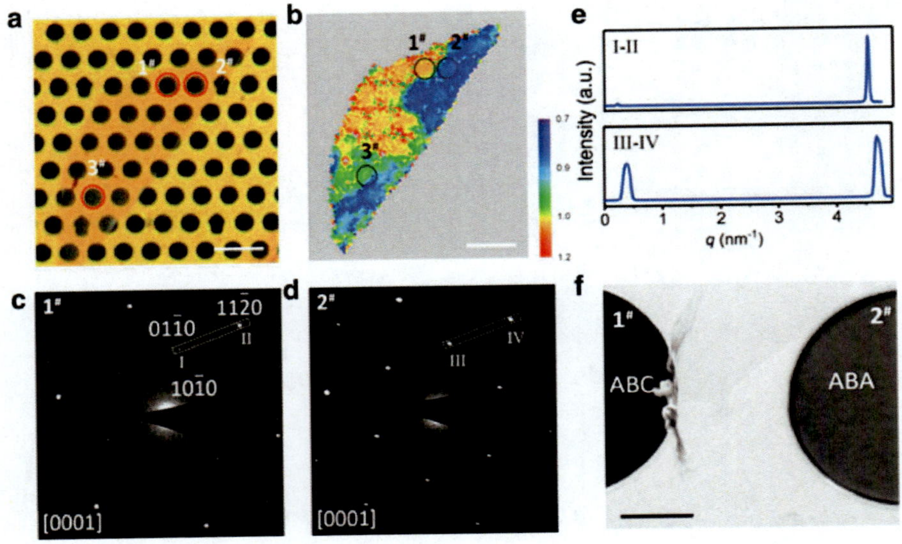

Fig. 1.21 **a** An optical micrograph (scale bar 10um) of 6 nm thick graphite flake suspended over a transmission electron microscope (TEM) grid. **b** Map of 2D peak of Raman spectrum of the flake (scale bar 10 μm) suspended over the TEM grid. **c, d** Selected-area electron diffraction imaging of suspended graphite flake at regions 1 and 2. **e** Intensity profile of the regions 1 and 2 along points I and II and III and IV, respectively. **f** Dark image of regions 1 and 2 obtained using first order $01\bar{1}0$ spot. Figure adapted from Yang et al. [85]

1.18.2 Transmission Electron Microscopy

Transmission electron microscope (TEM) selected-area electron diffraction imaging of graphite films as well as dark field imaging has been shown [84, 85] to give signatures of respective rhombohedral and Bernal phases of rhombohedral graphite as shown in Fig. 1.21.

A 6 nm thick graphite flake suspended on a TEM grid (Fig. 1.21a) with a corresponding map of FWHM of 2D peak of Raman spectrum (Fig. 1.21b) with red/yellow regions (Region 1) being the rhombohedral domain, with blue region (Region 2) being the Bernal hexagonal domain. Figure 1.21c, d show electron diffraction peaks on regions 1 and 2 respectively, with points $01\bar{1}0$ and $11\bar{2}0$ corresponding to first and second order diffraction peaks. Intensity profile of diffraction along points I to II and III to IV are shown on Fig. 1.21 and for rhombohedral stacked region first order diffraction peak intensity at $01\bar{1}0$ is minimal (i.e. close to zero) whereas the intensities from both regions are comparable for the case of Bernal hexagonal stacking. Hence, estimation of the ABC stacking fraction at a given region is enabled through study of the ratio intensity at the first order point $01\bar{1}0$ to second order point $11\bar{2}0$.

Using the above outlined property of two phases at first and second order diffraction points, by focusing on the first order diffraction peak $01\bar{1}0$ a dark field image of regions 1 and 2 were generated which is shown in Fig. 1.21f with a clear contrast

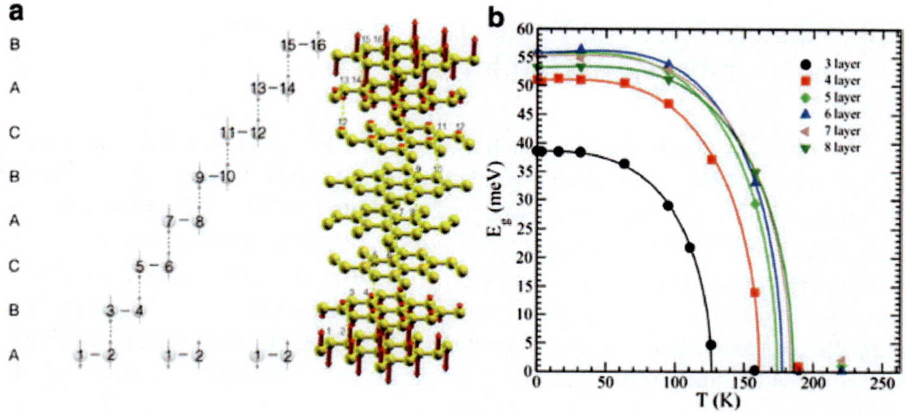

Fig. 1.22 **a** Crystal structure of rhombohedral graphite and illustrated configuration of spins at each sublattice with unequal magnitude. **b** Obtained magnetic gap as a function of temperature for layer numbers between three to eight. Figure adapted from Pamuk et al. [87]

difference between ABC and ABA stacked regions in agreement with the Raman 2D peak FWHM mapping.

1.19 Density Functional Theory Predicted Magnetic Gap

It has been found has been predicted and experimentally verified that the layer antiferromagnetic state highlighted in Sect. 1.13 dominates over other interaction induced states predicted by mean field calculations [32, 86]. Pamuk et al. [87], investigating rhombohedral stacks of graphene layers considered magnetic instabilities using density functional theory.

Their main finding is that layers of both surfaces of a rhombohedral stacked system are weakly ferrimagnetic individually (illustrated in Fig. 1.22a) arising from with opposite but unequal magnetic moments on atoms corresponding to each sublattice, however, leading to an overall antiferromagnetic configuration in the event of electronic coupling between the two surfaces. Layer numbers of up to 8 have been considered and it has been found that the gap size is maximum for 6-layers with an approximate value of around 55 meV, showing a slight decrease for layer numbers greater than 6 as shown in Fig. 1.22b. The gap size found for tri-layer systems, 38.6 meV is in a good agreement with transport experiments utilizing suspended tri-layer samples. A very interesting question the work raises is how the gap size would behave for even a greater number of layers where the electronic coupling between two surfaces could become absent. One may be able to probe the weakly ferrimagnetic nature of the two surfaces in transport experiments.

1.20 Angular Resolved Photoemission Spectroscopy of Rhombohedral Graphite

Angular resolved photoemission spectroscopy (ARPES) is one of the most direct methods of studying the electronic band structure of solids. It relies on the photoelectric effect and typically utilizes high energy photons (100–1000 eV) and allows the reconstruction of the band structure of a system through simultaneous detection of the energy of the incoming photons and the angle at which they leave the surface. The angle at which the electrons leave the surface allows a link between the energy of the photons and the conserved momentum on the plane of the sample. ARPES has been utilized to study band structure of graphene, see for example Bostwick et al. [59, 88], since the early days of its discovery.

ARPES experiments on rhombohedral graphene stacks had been carried on systems of up to 5–6 layers [89, 90]. The work which reports rhombohedral domains on grown multilayer graphene crystals of up to 6 layers by Pierucci et al. [90] is a growth experiment carried out on SiC, suggesting this system to be the optimal substrate for growth experiments on rhombohedral graphite domains. Higher quality results obtained on tri-layer systems have confirmed the simplest tight binding predictions for the band structure [44]. More recently, with the advances on identification of rhombohedral stacks of greater number of layers, an ARPES investigation was carried out on a 14-layer system for which the second derivative of the intensity maps obtained as well as the accompanied DFT calculations are shown in the Fig. 1.23 [91]. A dip of 25 meV size was found in the valance band around the K point towards the Γ and M points which has been attributed to the presence of the magnetic gap predicted by Pamuk et al. [87], however on a sample with a greater number of layers. It has been highlighted that electronic transport measurements are necessary to confirm the presence of the gap which will be the subject of this thesis.

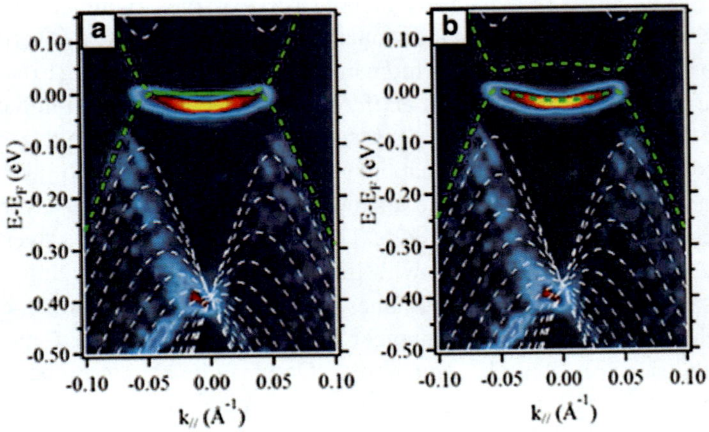

Fig. 1.23 Second derivative of an ARPES intensity map of a 14-layer system accompanied by DFT calculations **a** without a bad gap **b** with a spin polarized gap. Adapted from Henck et al. [91]

An ARPES experiment carried out more recently by Gao et al. [92] on ABC tri-layer graphene systems grown on strained copper substrates is very interesting from spectroscopy point of view, since it reports a gap on strained ABC tri-layer graphene as well as from growth point of view since it shows rhombohedral domains of up to 56% on CVD grown multilayer graphene. Hence, despite the thesis being about ABC stacked multilayer system beyond tri-layer, the work must be mentioned and is a future direction for growth experiments on multilayer systems of larger number of layers.

1.21 Shear Stress and Doping Induced Stacking Order Control

In the work reporting the directional assembly technique of assembling van der Waals heterostructures of rhombohedral graphite domains, for the devices reported in this thesis, it had been pointed out that shear force generated during exfoliation may play a role in the rhombohedral domains observed which are reported to be present in up to 29% of the exfoliated multilayers [85].

Furthermore, it was found that if the shear force applied during hBN encapsulation (which is necessary for high quality devices) was along armchair direction, stacking order alternates between ABA and ABC while a transition is absent for zigzag edge direction which is believed to be related to the 'stick–slip' displacement of graphene layers [93].

These experimental observations were further confirmed with density functional theory calculations of Nery et al. [94] that showed shear stress on multilayer graphene crystals induces reversible transitions from Bernal to rhombohedral structure, in agreement with Yang et al.'s [85] experimental observations and interpretations. The work reports calculations with regards to rhombohedral domains as a function angle to the armchair edge and the shear stress applied.

Lastly, a 'doping induced' stacking order transition evidenced by the powerful method of scanning near field optical microscopy in infra-red region was also reported very recently. In the work by Li et al. [95], doping dependence of stacking order in a tri-layer system was studied under extreme hole doping condition of above 6×10^{12} cm^{-2} on the bottom surface and an electron doping of above 7×10^{12} cm^{-2} on top surface a transition from ABA to ABC stacking was reported. The results are interpreted in terms of line/point defects associated with the stacking domain walls and an interplay of tensile and shear energies which is claimed to be controlled to a μeV/atom level.

1.22 Summary

In this introductory chapter, starting with the simplest theoretical description of graphene, we have reviewed existing literature on rhombohedral graphite, ranging from a few layers of ABC stacked graphene systems to what is expected for the case of bulk systems. Considering the most fundamental hopping parameters γ_0 and γ_1 it is shown that ABC stacked multilayers are a generalisation of graphene possessing Berry phase and chirality with respect to pseudospin degree of freedom. We have found that rhombohedral graphite consists of two 2D surfaces with a hexagonal Brillouin zone and undergoing a $E \sim p^N$ dispersion in the vicinity of K and K′ points, thus leading to, in principle, extremely flat surface bands.

We have established that once mapped on to a 1D chain rhombohedral graphite becomes identical to a Su–Schrieffer–Heeger chain. Considering the two band Hamiltonian of the bulk system we find that the gapless bulk has a Fermi surface of spiralling Dirac points, with the projection of it on the surface of the system leading to the 2D drumhead surface states thus making ABC graphite a nodal line semimetal whose flat band boundary are defined by the projection of bulk nodal line to the surface.

Assuming the simplest tight binding model with hopping parameters γ_0 and γ_1 are valid, we have investigated Landau level spectra of rhombohedral graphite systems of various thickness and established that at low energies Landau levels disperse as $B^{N/2}$ with B being the magnetic field. We have found that the extremely flat bands of rhombohedral graphite are potentially a host to correlated states such as anti-ferromagnetism and superconductivity as well as spontaneous quantum Hall states. As a unique layer dependent electron transport signature of rhombohedral graphite, we have looked-into single particle gap opening beyond a critical displacement field for various layer numbers.

Trigonal warping of the Fermi surface was found to be present after adaptation of the full tight binding framework that considers all parameters of Slonzewski-Weiss-McClure model. Remarkably, it was found that the presence of trigonal warping leads onto a finite number of Dirac points where conduction and valance bands touch, keeping the Berry phase of $N\pi$ present, that has been discussed in the absence hopping parameters γ_2, γ_2 and γ_4. We have reviewed the work on stacking faults in graphite systems ranging from tetra-layers to bulk. Lastly Raman and TEM characterisation of rhombohedral graphite systems as well as an ARPES assisted magnetic gap prediction and a doping induced stacking transition work was touched upon.

References

1. Klitzing KV, Dorda G, Pepper M (1980) New method for high-accuracy determination of the fine-structure constant based on quantized hall resistance. Phys Rev Lett 45(6):494–497. https://doi.org/10.1103/PhysRevLett.45.494
2. Thouless DJ, Kohmoto M, Nightingale MP, Den Nijs M (1982) Quantized hall conductance in a two-dimensional periodic potential. Phys Rev Lett 49(6):405–408. https://doi.org/10.1103/PhysRevLett.49.405
3. Haldane FDM (1988) Model for a quantum hall effect without landau levels: condensed-matter realization of the ‘parity anomaly.’ Phys Rev Lett 61(18):2015–2018. https://doi.org/10.1103/PhysRevLett.61.2015
4. Novoselov KS et al (2004) Electric field effect in atomically thin carbon films. Science 306(5696):666–669. https://doi.org/10.1038/nmat1849
5. Novoselov KS et al (2005) Two-dimensional gas of massless Dirac fermions in graphene. Nature 438(7065):197–200. https://doi.org/10.1038/nature04233

6. Zhang Y, Tan YW, Stormer HL, Kim P (2005) Experimental observation of the quantum Hall effect and Berry's phase in graphene. Nature 438(7065):201–204. https://doi.org/10.1038/nature04235
7. Geim AK, Novoselov KS (2007) The rise of graphene. Nat Mater 6(3):183–191. https://doi.org/10.1038/nmat1849
8. Kane CL, Mele EJ (2005) Quantum Spin hall effect in graphene. Phys Rev Lett 95(22):1–4. https://doi.org/10.1103/PhysRevLett.95.226801
9. Kane CL, Mele EJ (2005) Z2 topological order and the quantum spin hall effect. Phys Rev Lett 95(14):3–6. https://doi.org/10.1103/PhysRevLett.95.146802
10. Moore JE, Balents L (2007) Topological invariants of time-reversal-invariant band structures. Phys Rev B - Condens Matter Mater Phys 75(12):3–6. https://doi.org/10.1103/PhysRevB.75.121306
11. Fu L, Kane CL, Mele EJ (2007) Topological insulators in three dimensions. Phys Rev Lett 98(10):1–4. https://doi.org/10.1103/PhysRevLett.98.106803
12. Roy R (2009) Topological phases and the quantum spin Hall effect in three dimensions. Phys Rev B - Condens Matter Mater Phys 79(19):1–5. https://doi.org/10.1103/PhysRevB.79.195322
13. Hsieh D et al (2008) A topological Dirac insulator in a quantum spin Hall phase. Nature 452(7190):970–974. https://doi.org/10.1038/nature06843
14. Hsieh D et al (2009) A tunable topological insulator in the spin helical Dirac transport regime. Nature 460(7259):1101–1105. https://doi.org/10.1038/nature08234
15. Armitage NP, Mele EJ, Vishwanath A (2018) Weyl and Dirac semimetals in three-dimensional solids. Rev Mod Phys 90(1):15001. https://doi.org/10.1103/RevModPhys.90.015001
16. Novoselov KS et al (2005) Two-dimensional atomic crystals. Proc Natl Acad Sci USA 102(30):10451–10453. https://doi.org/10.1073/pnas.0502848102
17. Dean CR et al (2010) Boron nitride substrates for high-quality graphene electronics. Nat Nanotechnol 5(10):722–726. https://doi.org/10.1038/nnano.2010.172
18. Ponomarenko LA et al (2013) Cloning of Dirac fermions in graphene superlattices. Nature 497(7451):594–597. https://doi.org/10.1038/nature12187
19. Dean CR et al (2013) Hofstadter's butterfly and the fractal quantum Hall effect in moiré superlattices. Nature 497(7451):598–602. https://doi.org/10.1038/nature12186
20. Hunt B et al (2013) Massive Dirac Fermions and Hofstadter Butterfly in a van der Waals Heterostructure. Science 340(6139):1427–1430. https://doi.org/10.1126/science.1237240
21. Gorbachev RV et al (2014) Detecting topological currents in graphene superlattices. Science 346(6208):448–451. https://doi.org/10.1126/science.1254966
22. Novoselov KS, Mishchenko A, Carvalho A, Castro Neto AH (2016) 2D materials and van der Waals heterostructures. Science 353(6298):aac9439. https://doi.org/10.1126/science.aac9439
23. Bandurin DA et al (2017) High electron mobility, quantum Hall effect and anomalous optical response in atomically thin InSe. Nat Nanotechnol 12(3):223–227. https://doi.org/10.1038/nnano.2016.242
24. Guinea F, Castro Neto AH, Peres NMR (2006) Electronic states and Landau levels in graphene stacks. Phys Rev B 73(24):245426. https://doi.org/10.1103/PhysRevB.73.245426
25. Arovas DP, Guinea F (2008) Stacking faults, bound states, and quantum Hall plateaus in crystalline graphite. Phys Rev B - Condens Matter Mater Phys 78(24):1–18. https://doi.org/10.1103/PhysRevB.78.245416
26. Min H, MacDonald AH (2008) Chiral decomposition in the electronic structure of graphene multilayers. Phys Rev B - Condens Matter Mater Phys 77(15):1–5. https://doi.org/10.1103/PhysRevB.77.155416
27. Min H, MacDonald AH (2008) Electronic structure of multilayer graphene. Prog Theor Phys Suppl 176(176):25. https://doi.org/10.1143/PTPS.176.227
28. Neto AHC, Guinea F, Peres NMR, Novoselov KS, Geim AK (2009) The electronic properties of grapheme. Rev Mod Phys 81. https://doi.org/10.1103/RevModPhys.81.109
29. Koshino M, McCann E (2009) Trigonal warping and Berry's phase N-pi in ABC-stacked multilayer grapheme. Phys Rev B - Condens Matter Mater Phys 80(16):1–8. https://doi.org/10.1103/PhysRevB.80.165409

30. Zhang F, Sahu B, Min H, MacDonald AH (2010) Band structure of ABC-stacked graphene trilayers. Phys Rev B - Condens Matter Mater Phys 82(3):1–10. https://doi.org/10.1103/PhysRevB.82.035409
31. Ho CH, Chang CP, Lin MF (2016) Evolution and dimensional crossover from the bulk subbands in ABC-stacked graphene to a three-dimensional Dirac cone structure in rhombohedral graphite. Phys Rev B 93(7). https://doi.org/10.1103/PhysRevB.93.075437
32. Zhang F, Jung J, Fiete GA, Niu Q, MacDonald AH (2011) Spontaneous quantum hall states in chirally stacked few-layer graphene systems. Phys Rev Lett 106(15):1–4. https://doi.org/10.1103/PhysRevLett.106.156801
33. Kopnin NB, Heikkilä TT, Volovik GE (2011) High-temperature surface superconductivity in topological flat-band systems. Phys Rev B - Condens Matter Mater Phys 83(22): 1–4. https://doi.org/10.1103/PhysRevB.83.220503
34. Xiao R, Tasnádi F, Koepernik K, Venderbos JWF, Richter M, Taut M (2011) Density functional investigation of rhombohedral stacks of graphene: topological surface states, nonlinear dielectric response, and bulk limit. Phys Rev B - Condens Matter Mater Phys 84(16):1–15. https://doi.org/10.1103/PhysRevB.84.165404
35. Heikkilä TT, Volovik GE (2011) Dimensional crossover in topological matter: evolution of the multiple Dirac point in the layered system to the flat band on the surface. JETP Lett 93(2):59–65. https://doi.org/10.1134/S002136401102007X
36. Burkov AA, Hook MD, Balents L (2011) Topological nodal semimetals. Phys Rev B - Condens Matter Mater Phys 84(23):1–14. https://doi.org/10.1103/PhysRevB.84.235126
37. Myhro K et al (2018) Large tunable intrinsic gap in rhombohedral-stacked tetralayer graphene at half filling. 2D Mater 5(4). https://doi.org/10.1088/2053-1583/aad2f2
38. McCann E (2012) Electronic Properties of Monolayer and Bilayer Graphene. In: Raza H (ed) Graphene nanoelectronics: metrology, synthesis, properties and applications. Springer, Berlin Heidelberg, pp 237–275
39. Bernal JD (1924) The structure of graphite. Proc R Soc London Ser A Contain Pap Math Phys Character 106(740):749–773. https://doi.org/10.1098/rspa.1924.0101
40. Lipson H, Stokes A (1942) The structure of graphite. Proc R Soc London Ser A Math Phys Sci 181(984):101–105. https://doi.org/10.1098/rspa.1942.0063
41. Wallace PR (1947) The band theory of graphite. Phys Rev 71(9):622–634. https://doi.org/10.1103/PhysRev.71.622
42. Kuzmenko AB, Crassee I, Van Der Marel D, Blake P, Novoselov KS (2009) Determination of the gate-tunable band gap and tight-binding parameters in bilayer graphene using infrared spectroscopy. Phys Rev B - Condens Matter Mater Phys 80(16):1–12. https://doi.org/10.1103/PhysRevB.80.165406
43. Mak KF, Shan J, Heinz TF (2010) Electronic structure of few-layer graphene: Experimental demonstration of strong dependence on stacking sequence. Phys Rev Lett 104(17):1–4. https://doi.org/10.1103/PhysRevLett.104.176404
44. Bao C et al (2017) Stacking-dependent electronic structure of Trilayer graphene resolved by nanospot angle-resolved photoemission spectroscopy. Nano Lett 17(3):1564–1568. https://doi.org/10.1021/acs.nanolett.6b04698
45. Berry MV (1984) Quantal phase factors accompanying adiabatic changes. Proc R Soc A Math Phys Eng Sci 392(1802):45–57. https://doi.org/10.1098/rspa.1984.0023
46. Novoselov KS et al (2006) Unconventional quantum Hall effect and Berry's phase of 2π in bilayer graphene. Nat Phys 2(3):177–180. https://doi.org/10.1038/nphys245
47. Slizovskiy S, McCann E, Koshino M, Fal'ko VI (2019) Films of rhombohedral graphite as two-dimensional topological semimetals. Commun Phys 2(1):1–10. https://doi.org/10.1038/s42005-019-0268-8
48. Gu ZC, Wen XG (2009) Tensor-entanglement-filtering renormalization approach and symmetry-protected topological order. Phys Rev B - Condens Matter Mater Phys 80(15):1–23. https://doi.org/10.1103/PhysRevB.80.155131
49. Chen X, Gu ZC, Liu ZX, Wen XG (2013) Symmetry protected topological orders and the group cohomology of their symmetry group. Phys Rev B - Condens Matter Mater Phys 87(15):1–48. https://doi.org/10.1103/PhysRevB.87.155114

50. Zak J (1989) Berrys phase for energy bands in solids. Phys Rev Lett 62(23):2747–2750. https://doi.org/10.1103/PhysRevLett.62.2747
51. Asbóth JK, Oroszlány L, Pályi A (2016) The Su-Schrieffer-Heeger (SSH) model. In: A short course on topological insulators: band structure and edge states in one and two dimensions. Springer International Publishing, Cham pp 1–22
52. Henni Y et al (2016) Rhombohedral multilayer graphene: a magneto-Raman scattering study. Nano Lett 16(6):3710–3716. https://doi.org/10.1021/acs.nanolett.6b01041
53. Sena SHR, Pereira JM, Farias GA, Peeters FM (2012) Cyclotron resonance of trilayer grapheme. Phys Rev B - Condens Matter Mater Phys 86(8):1–9. https://doi.org/10.1103/PhysRevB.86.085412
54. Min H, Borghi G, Polini M, MacDonald AH (2008) Pseudospin magnetism in grapheme. Phys Rev B - Condens Matter Mater Phys 77(4):2–5. https://doi.org/10.1103/PhysRevB.77.041407
55. Nandkishore R, Levitov L (2010) Dynamical screening and excitonic instability in bilayer graphene. Phys Rev Lett 104(15):3–6. https://doi.org/10.1103/PhysRevLett.104.156803
56. Zhang F, Min H, Polini M, MacDonald AH (2010) Spontaneous inversion symmetry breaking in graphene bilayers. Phys Rev B - Condens Matter Mater Phys 81(4):2–5. https://doi.org/10.1103/PhysRevB.81.041402
57. Mayorov AS et al (2011) Interaction-driven spectrum reconstrunction in bilayer graphene. Science 333(August):860–864
58. Koshino M (2010) Interlayer screening effect in graphene multilayers with ABA and ABC stacking. Phys Rev B - Condens Matter Mater Phys 81(12):1–7. https://doi.org/10.1103/PhysRevB.81.125304
59. Ohta T, Bostwick A, Seyller T, Horn K, Rotenberg E (2006) Controlling the electronic structure of bilayer graphene. Science 313(5789):951–954. https://doi.org/10.1126/science.1130681
60. Castro EV et al (2007) Biased bilayer graphene: semiconductor with a gap tunable by the electric field effect. Phys Rev Lett 99(21):8–11. https://doi.org/10.1103/PhysRevLett.99.216802
61. Oostinga JB, Heersche HB, Liu X, Morpurgo AF, Vandersypen LMK (2008) Gate-induced insulating state in bilayer graphene devices. Nat Mater 7(2):151–157. https://doi.org/10.1038/nmat2082
62. Zhang Y et al (2009) Direct observation of a widely tunable bandgap in bilayer graphene. Nature 459(7248):820–823. https://doi.org/10.1038/nature08105
63. Mak KF, Lui CH, Shan J, Heinz TF (2009) Observation of an electric-field-induced band gap in bilayer graphene by infrared spectroscopy. Phys Rev Lett 102(25):100–103. https://doi.org/10.1103/PhysRevLett.102.256405
64. Lui CH, Li Z, Mak KF, Cappelluti E, Heinz TF (2011) Observation of an electrically tunable band gap in trilayer graphene. Nat Phys 7(12):944–947. https://doi.org/10.1038/nphys2102
65. Zou K, Zhang F, Clapp C, MacDonald AH, Zhu J (2013) Transport studies of dual-gated ABC and ABA trilayer graphene: band gap opening and band structure tuning in very large perpendicular electric fields. Nano Lett 13(2):369–373. https://doi.org/10.1021/nl303375a
66. Kopnin NB, Ijäs M, Harju A, Heikkilä TT (2013) High-temperature surface superconductivity in rhombohedral graphite. Phys Rev B - Condens Matter Mater Phys 87(14):1–4. https://doi.org/10.1103/PhysRevB.87.140503
67. Kittel C (1987) Quantum theory of solids. Wiley, New York
68. Cao Y et al (2018) Unconventional superconductivity in magic-angle graphene superlattices. Nature 556(7699):43–50. https://doi.org/10.1038/nature26160
69. Yankowitz M et al (2019) Tuning superconductivity in twisted bilayer graphene. Science 363(6431):1059–1064. https://doi.org/10.1126/science.aav1910
70. Lu X et al (2019) Superconductors, orbital magnets and correlated states in magic-angle bilayer graphene. Nature 574(7780):653–657. https://doi.org/10.1038/s41586-019-1695-0
71. Latil S, Henrard L (2006) Charge carriers in few-layer graphene films. Phys Rev Lett 97(3):1–4. https://doi.org/10.1103/PhysRevLett.97.036803
72. Aoki M, Amawashi H (2007) Dependence of band structures on stacking and field in layered graphene. Solid State Commun 142(3):123–127. https://doi.org/10.1016/j.ssc.2007.02.013
73. Min H, MacDonald AH (2008) Electronic structure of multilayer graphene. Prog Theor Phys Suppl 176:227–252. https://doi.org/10.1143/PTPS.176.227

74. Koshino M, McCann E (2013) Multilayer graphenes with mixed stacking structure: Interplay of Bernal and rhombohedral stacking. Phys Rev B - Condens Matter Mater Phys 87(4):1–12. https://doi.org/10.1103/PhysRevB.87.045420
75. Taut M, Koepernik K, Richter M (2014) Electronic structure of stacking faults in rhombohedral graphite. Phys Rev B - Condens Matter Mater Phys 90:1–8. https://doi.org/10.1103/PhysRevB.90.085312
76. Craciun MF, Russo S, Yamamoto M, Oostinga JB, Morpurgo AF, Tarucha S (2009) Trilayer graphene is a semimetal with a gate-tunable band overlap. Nat Nanotechnol 4(6):383–388. https://doi.org/10.1038/nnano.2009.89
77. Nery JP, Calandra M, Mauri F (2021) Ab-initio energetics of graphite and multilayer graphene : stability of Bernal versus rhombohedral stacking. 2D Materials 8:035006. https://doi.org/10.1088/2053-1583/abec23
78. Lui CH, Li Z, Chen Z, Klimov PV, Brus LE, Heinz TF (2011) Imaging stacking order in few-layer graphene. Nano Lett 11(1):164–169. https://doi.org/10.1021/nl1032827
79. Lui CH et al (2012) Observation of layer-breathing mode vibrations in few-layer graphene through combination raman scattering. Nano Lett 12(11):5539–5544. https://doi.org/10.1021/nl302450s
80. Kim DS et al (2015) Stacking structures of few-layer graphene revealed by phase-sensitive infrared nanoscopy. ACS Nano 9(7):6765–6773. https://doi.org/10.1021/acsnano.5b02813
81. Jorio A, Cançado LG, Malard LM (2017) Vibrations in grapheme. 2D Mater Prop Devices 71–89. https://doi.org/10.1017/9781316681619.006
82. Herziger F et al (2014) Two-dimensional analysis of the double-resonant 2D Raman mode in bilayer graphene. Phys Rev Lett 113(18):1–5. https://doi.org/10.1103/PhysRevLett.113.187401
83. Torche A, Mauri F, Charlier J-C, Calandra M (2017) First-principles determination of the Raman fingerprint of rhombohedral graphite. Phys Rev Mater 1(4):041001. https://doi.org/10.1103/PhysRevMaterials.1.041001
84. Latychevskaia T et al (2019) Stacking transition in rhombohedral graphite. Front Phys 14:13608. https://doi.org/10.1007/s11467-018-0867-y
85. Yang Y et al (2019) Stacking order in graphite films controlled by van der Waals technology. Nano Lett 19:8526–8532. https://doi.org/10.1021/acs.nanolett.9b03014
86. Lee Y et al (2014) Competition between spontaneous symmetry breaking and single-particle gaps in trilayer graphene. Nat Commun 5:5656. https://doi.org/10.1038/ncomms6656
87. Pamuk B, Baima J, Mauri F, Calandra M (2017) Magnetic gap opening in rhombohedral-stacked multilayer graphene from first principles. Phys Rev B 95(7). https://doi.org/10.1103/PhysRevB.95.075422
88. Bostwick A, Ohta T, Seyller T, Horn K, Rotenberg E (2007) Quasiparticle dynamics in graphene. Nat Phys 3(1):36–40. https://doi.org/10.1038/nphys477
89. Coletti C et al (2013) Revealing the electronic band structure of trilayer graphene on SiC: an angle-resolved photoemission study. Phys Rev B - Condens Matter Mater Phys 88(15):1–6. https://doi.org/10.1103/PhysRevB.88.155439
90. Pierucci D et al (2015) Evidence for flat bands near the Fermi level in epitaxial rhombohedral multilayer graphene. ACS Nano 9(5):5432–5439. https://doi.org/10.1021/acsnano.5b01239
91. Henck H et al (2018) Flat electronic bands in long sequences of rhombohedral-stacked graphene. Phys Rev B 97(24):1–6. https://doi.org/10.1103/PhysRevB.97.245421
92. Gao Z et al (2020) Large area epitaxial growth of curvature-stabilized ABC trilayer graphene. Nat Communs 11:1–10. https://doi.org/10.1038/s41467-019-14022-3
93. Dienwiebel M et al (2004) Superlubricity of graphite. Phys Rev Lett 92(12):126101. https://doi.org/10.1103/PhysRevLett.92.126101
94. Nery JP, Calandra M, Mauri F (2020) Long-range rhombohedral stacked graphene through shear. Nano Lett 20(7):5017–5023. https://doi.org/10.1021/acs.nanolett.0c01146
95. Li H et al (2020) Global control of stacking order phase transition by doping and electric field in a few-layer graphene. Nano Lett 20(5):310603112. https://doi.org/10.1021/acs.nanolett.9b05092

Chapter 2
Fundamentals of Electron Transport

2.1 Introduction

In this chapter a general introduction and overview of electronic transport concepts such as effective mass and mobility will be given where Ohms law will be derived through both from a classical picture as well as a semi-classical picture. It will be highlighted that from a semi-classical point of view low temperature conductivity of metals is a Fermi surface property. Further to this, the diffusive viewpoint of charge transport which holds at low temperatures will be outlined. Hall effect, which is a very fundamental phenomenon to transport studies will be introduced. Higher magnetic fields will be considered where Shubnikov-de Haas oscillations will be introduced as well as the quantum Hall effect. Anomalous Hall effects of both magnetic, spin and valley (both intrinsic and extrinsic) origins will be introduced as well as their quantised versions which take place at zero magnetic fields. Lastly, low temperature, quantum transport phenomena of weak (anti)localization will be explained in a pedagogical way, which will also become relevant at latter sections reporting the experimental results.

2.2 Effective Mass, Mobility, and Electrical Conductivity

To consider dynamics of electrons making up the crystal classically, one must consider superposition of electron wavefunctions in plane wave states as wave packets for which the motion resembles that of a classical particle. Assuming plane wave solutions in a two-dimensional space for illustrative purposes, we can calculate the group velocity of an electron wave packet as

$$v = \frac{d\omega}{d\mathbf{k}} = \frac{1}{\hbar}\frac{d\varepsilon}{d\mathbf{k}} \tag{2.1}$$

S. Ozdemir, *Electronic Properties of Rhombohedral Graphite*, Springer Theses,
https://doi.org/10.1007/978-3-030-88307-2_2

with a net electric field **E** is applied to the crystal, where the wave packet moves with its total energy remaining constant, the change in its kinetic energy is equal to

$$\delta\varepsilon = -e\mathbf{E}\mathbf{v}\delta t \tag{2.2}$$

at a time interval of δt. We know from Eq. (2.1) that $\delta\varepsilon = \boldsymbol{v}\hbar\delta\boldsymbol{k}$ which would then lead to the expression

$$\hbar\frac{d\mathbf{k}}{dt} = e\mathbf{E} \tag{2.3}$$

with $\hbar\mathbf{k}$ being referred to as crystal momentum of an electron. Taking a time derivative of Eq. (2.3) and combining it with Eq. (2.2) one obtains the equation

$$m_e\frac{d\mathbf{v}}{dt} = -e\mathbf{E} \tag{2.4}$$

where $m_e = \hbar^2/\left(d^2\varepsilon/d\boldsymbol{k}^2\right)$ is the effective mass arising from the band curvature [1, 2]. We note that the concept of such an effective mass does not apply to graphene, and in graphene instead of second derivative of dispersion relation which diverges, the relativistic equation $E = mc_{eff}^2$ is used away from the Dirac point [3], with speed of light being replaced by the Fermi velocity. In bilayer graphene, which has massive chiral quasiparticles arising from quadratic dispersion such an approximation is valid. The case of ABC stacked multilayer graphene; is more complicated with the effective mass approximation being only valid for the case of an induced quadratic dispersion when additional tight binding hopping terms are considered.

Having established electrons having some form of effective mass we can move to Drude model of electron transport and prove how Ohms law arises because of it. Equation (2.4) on its own would imply infinitely accelerating electrons in the presence of a constant electric field, however, we know that this not the case as in practice as electrons are scattered because of collisions (except ballistic systems). The possibility of collisions was first considered by Paul Drude in 1900 where he used an additional term leading to an exponential decay of electron velocity in the removal of an applied electric field as shown below

$$m_e\frac{d\mathbf{v}}{dt} = -e\mathbf{E} - m_e\frac{\mathbf{v}}{\tau} \tag{2.5}$$

with τ being the time constant [4]. In the presence of a finite electric field the equation has a steady state solution, which is referred to as the drift velocity which is shown in equation below

$$\mathbf{v}_d = -\frac{e\tau}{m_e}\mathbf{E} = \mu_e\mathbf{E} \tag{2.6}$$

with τ time constant being the time for which the electrons travel before getting scattered. The constant of proportionality between the drift velocity and the applied electric field, depicted by μ_e is known as the electron mobility. Having established the drift velocity, we can re-write the equation for current density, ($\boldsymbol{j} = ne\boldsymbol{v_d}$), as

$$\mathbf{j} = \frac{ne^2\tau}{m_e}\mathbf{E} = \sigma\mathbf{E} \tag{2.7}$$

which is the Ohms law with n being the carrier density and σ being the conductivity.

So far, we have paid attention to wave packets that have negative charges, in addition to electron wave packets, there are also what we call hole wave packets that arise with a positive charge in unoccupied states of the valance band. Combining electrons of conduction band and the holes of valance band one can write the equation for resistivity/conductivity of a two-band system (like thin film rhombohedral graphite in low energy limit) as

$$\frac{1}{\rho} = \sigma = ne\mu_e + pe\mu_h \tag{2.8}$$

where n is the concentration of electrons and p is the concentration of holes.

2.3 Ohms Law Using Boltzmann Transport Equation

You may note that in previous section we have started with quantum mechanics of the crystal, deriving a crystal momentum and introducing an effective mass (note the use of the word effective), but then used a very classical equation, as Drude did before the development of the necessary quantum theory, to prove the Ohm's Law and the fundamental transport concept of electron mobility. When we obtained Eq. (2.4), for the momentum introduced, we have used the definition of crystal momentum, rather than an electron momentum. This is because when an electric field is applied to a crystal, a momentum is generated on the electron as well as the lattice of the crystal. It was after the development of Drude model in 1900, that quantum mechanics was developed, with significant implications being made to solid state physics through the Bloch theorem [5], which states that electrons are not free particles, but are subject to a periodic potential arising from the crystal lattice. Thus, in a way, we have a problem to solve. We will therefore use what is called a Boltzmann transport equation to obtain the same expression for conductivity that we obtained from Drude model and meanwhile we will take care of the necessary quantum mechanics.

In the presence of an electric field, E, according to Boltzmann transport theory, a steady system is described with a function $f(\boldsymbol{r}, \boldsymbol{k}, t)$ which restores to a Fermi-Dirac distribution

$$f_0(\mathbf{k}) = f(\mathbf{r}, \mathbf{k}, t)|_{E=0} = \frac{1}{e^{\beta(E(\mathbf{k})-E_F)} + 1} \tag{2.9}$$

in the presence of zero electric field.

Assuming there are no scattering events, one can write the following equation for a system, before and after a time frame of dt

$$f(\mathbf{r}, \mathbf{k}, t) = f(\mathbf{r} + \mathbf{v}dt, \mathbf{k} - (e\mathbf{E}/\hbar)dt, t + dt) \tag{2.10}$$

given that the number of quasiparticles are indeed conserved. In the presence of scattering events however, such as from phonons or defects and impurities, the equality above does not hold. To take into account such scattering effects within a time frame of dt, we include an additional term

$$f(\mathbf{r}, \mathbf{k}, t) + \left(\frac{\partial f}{\partial t}\right)_S dt = f(\mathbf{r} + \mathbf{v}dt, \mathbf{k} - (e\mathbf{E}/\hbar)dt, t + dt) \tag{2.11}$$

with the Taylor expansion of the above expression, assuming that time scale dt is infinitesimally small, yielding the following expression

$$\left(\frac{\partial f}{\partial t}\right)_S = \mathbf{v} \cdot \nabla_{\mathbf{r}} f - e\mathbf{E}\frac{\nabla_{\mathbf{k}} f}{\hbar} + \frac{\partial f}{\partial t} \tag{2.12}$$

which is known as the Boltzmann transport equation. Lets consider a simple expression for the scattering term, in the limit when the applied force that drove our system to $f(\boldsymbol{k})$ from $f_0(\boldsymbol{k})$ becomes absent,

$$\left(\frac{\partial f}{\partial t}\right)_S = \frac{-f - f_0}{\tau} \tag{2.13}$$

where we modelled our system to exponentially relax back to the equilibrium state a time constant of τ, i.e. integration of Eq. (2.13) yields the expression

$$(f - f_0)_t = (f - f_0)_{t=0} e^{-t/\tau} \tag{2.14}$$

in this point we note that we have used the same approximation in the Drude model.

We shall now consider the system in a steady state $\left(\frac{\partial f}{\partial t} = 0\right)$ after the application of a time-independent electric field $\mathbf{E} = E_0\hat{x}$. Furthermore we shall assume that our system is spatially homonogenous which means that $\nabla_r f = 0$ resulting in the Boltmann transport equation for the system to becomes

$$f(\mathbf{k}) = f_0(\mathbf{k}) + \frac{e}{\hbar}\tau\mathbf{E}\cdot\nabla_{\mathbf{k}} f(\mathbf{k}) \tag{2.15}$$

which in the small **E** limit becomes

$$f(\mathbf{k}) \cong f_0(\mathbf{k}) + \frac{e}{\hbar}\tau(\mathbf{k})\mathbf{E}\cdot\nabla_{\mathbf{k}} f_0(\mathbf{k}) \tag{2.16}$$

and for an $\mathbf{E} = E_0\hat{x}$, the expression above could be re-written as

$$f(\mathbf{k}) \cong f_0\left(\mathbf{k} + \frac{e}{\hbar}\tau(\mathbf{k})\mathbf{E}\right) \tag{2.17}$$

which is known as the linear Boltzmann equation. We can see that the effect of the electric field is to shift the state described by the Fermi-distribution function from a momentum of $\boldsymbol{k}$ to a momentum of $\boldsymbol{k} + \frac{e}{\hbar}\tau(\boldsymbol{k})E_x$, thus shifting the overall Fermi surface in the case of a 3D system in kx direction and resulting in two misaligned quasi Fermi-levels in 2D case at positive and negative momenta [6]. Now, using a 3D system as an illustration example (despite the thin film nature of rhombohedral graphite films studied in this thesis we will see that 3D effects are also observed) we will derive an expression for current density within Boltzman transport equation framework.

The equation for current density is given

$$j_x \cong -2e\int \frac{d^3k}{(2\pi)^3} v(\mathbf{k})\cdot\left(f_0(\mathbf{k}) + \frac{e\tau}{\hbar}\mathbf{E}\frac{\partial f_0(\mathbf{k})}{\partial k_x}\right) \tag{2.18}$$

The first term in the bracket vanishes as $v(-\boldsymbol{k}) = v(\boldsymbol{k})$, hence leading to the equation

$$j_x \cong -2e\int \frac{d^3k}{(2\pi)^3} v_x \frac{e\tau}{\hbar}\mathbf{E}\frac{\partial f_0}{\partial k_x} \tag{2.19}$$

we can we write $\frac{\partial f_0(\boldsymbol{k})}{\partial k_x}$ as $\frac{\partial f_0(\boldsymbol{k})}{\partial \varepsilon}\frac{\partial \varepsilon}{\partial k_x} = \frac{\partial f_0(\boldsymbol{k})}{\partial \varepsilon}\hbar v_x$, which would lead to the equation for current density becoming

$$j_x = -\frac{2e^2\hbar\mathbf{E}}{8\pi^3}\int d^3k v_x^2\tau\frac{\partial f_0}{\partial \varepsilon} \tag{2.20}$$

we know that at $T \cong 0K$ the function $\frac{\partial f_0}{\partial \varepsilon}$ approaches to being a Dirac delta function, which leads to

$$j_x = -\frac{2e^2\hbar\mathbf{E}}{8\pi^3}\int d^3\mathbf{k} v_x^2\delta(E - E_F) \tag{2.21}$$

at this point we carry out the expansion $d^3k = d\mathbf{S}d\mathbf{k} = d\mathbf{S}\frac{d\varepsilon}{\hbar^2 v(\mathbf{k})}$ and modifying the equation for current density as

$$j_x \cong -\frac{2e^2\hbar\mathbf{E}}{8\pi^3\hbar}\int d\mathbf{S}d\varepsilon v_x^2/v(\mathbf{k})\delta(E-E_F) \tag{2.22}$$

and obtaining

$$j_x \cong -\frac{2e^2\mathbf{E}}{8\pi^3\hbar}\int d\mathbf{S_F}\frac{v_x^2\tau}{v_F} \tag{2.23}$$

where it has to be said at this point that contribution to current can clearly be seen to be only at the Fermi surface. We can approximate $v_F \cong \sqrt{3}v_x$ which would lead to

$$j_x \cong -\frac{2e^2\mathbf{E}}{8\pi^3\hbar}\frac{4}{3}\pi k_F^2 v_F\tau = -\frac{2e^2\mathbf{E}\tau}{8\pi^3 m_e}\frac{4}{3}\pi k_F^3 \tag{2.24}$$

we know that $n = k_F^3/3\pi^2$ which results in the Ohms Law

$$j_x = \frac{ne^2\tau}{m_e}\mathbf{E} = \sigma\mathbf{E} \tag{2.25}$$

which means that we have solved a transport problem semi-classically using the Fermi-Dirac distribution. One can argue that this the more legitimate derivation as we now know that not all electrons in a metal participate in conduction process at low temperatures which is something we assumed in the Drude model.

An interesting question which naturally arises is, do electrons move with the drift velocity or Fermi velocity. In light of the derivation we have carried out, one can modify classical equation for current density as

$$j = e\left[n\frac{v_d}{v_F}\right]v_F \tag{2.26}$$

which means that only a certain fraction of electrons near the Fermi surface move with a Fermi velocity rather than the entire electrons moving with a drift velocity [6].

2.4 Quasi Fermi Level Separation and Diffusive Electron Transport Viewpoint

We have, in Sect. 1, assumed that with the application of an electric field E, all electrons of the system under consideration move with a drift velocity v_d, and we have linked this drift velocity to the a characteristic quantity, namely the mobility, which is the drift velocity charges gain per unit electric field. In section two however, we have seen that, taking care of necessary quantum mechanics, at low temperatures, the conductance is in fact a Fermi surface property. Despite having written the classical formula for current density in an alternative way in Eq. (2.26), we did not carry out a proper derivation.

One can visualise the shift in $f(\mathbf{k})$ as a result of application of a bias voltage across a mesoscopic sample (see Fig. 2.1a), and generation of net electric field (Fig. 2.1b), as formation of two quasi Fermi levels (Fig. 2.1c), one for the electrons moving in the direction of the field applied, which are the states occupying positive momentum values in this instance, which is depicted as F^+ in Fig. 2.1c, and one for the states occupying negative momentum values, which is depicted as F^-. One can see that, the states that carry current are only the ones that shift from negative momentum values to the positive momentum values, as a result of application of the net electric field. Defining the electric field induced momentum change as $k_1 = \frac{e}{\hbar}\tau(\boldsymbol{k})E_x$, we can write the separation between two quasi fermi levels as

$$\Delta E = \frac{2\hbar^2 k_F k_1}{m_e} = 2e\tau v_F E_x = 2eE_x L_m \tag{2.27}$$

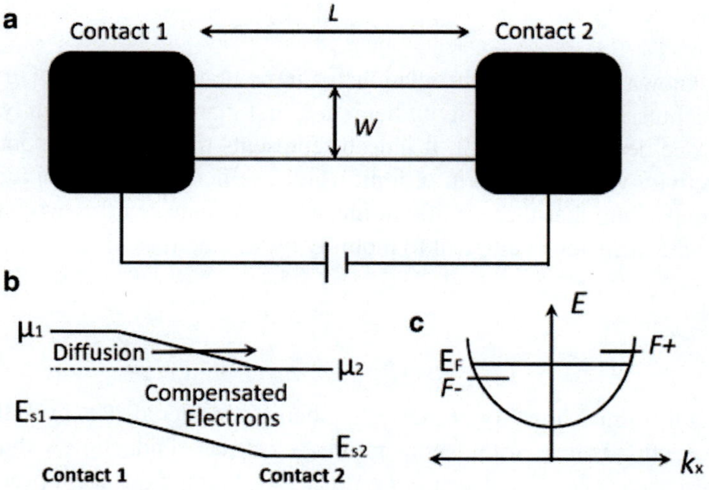

Fig. 2.1 **a** Potential difference applied across two contracts on a strip of width W and length L. **b** Chemical potential imbalance ($\mu_1 - \mu_2$) around the contacts leading to electron diffusion. **c** Illustration of conductivity as a Fermi surface property where compensated electrons away from Fermi energy do not take part.

where we have defined the quantity mean free path, L_m. It can be seen that the shift in energy levels as an electric field of E_x is applied is proportional to the amount of energy electrons gain within time frame of a mean fee path.

One can see on Fig. 2.1b, that as a result of application of a net electric field, a chemical potential gradient is generated across the sample, with electrons below an energy of μ_2, not contributing to the current, in other words being compensated (we are assuming to be at very low temperatures). Defining an average Fermi level, $E_f = \left(F^+ + F^-\right)/2$ one can define a Fermi surface, and associated density of states, $\frac{dn}{dE_F}$, which means, in the vicinity of contact 1 (Fig. 2.1a), we have an electron density of

$$n_e = \frac{dn}{dE_F}(\mu_1 - \mu_2) \tag{2.28}$$

in the regime where transport will happen, whereas in the vicinity of contact 2 all electrons are compensated. This creates a diffusion gradient of electrons, which can be modelled by Ficks law

$$\mathbf{j} = eD\frac{\mu_1 - \mu_2}{L}\frac{dn}{dE_F}\hat{x} = e^2 D\frac{dn}{dE_F}\mathbf{E} = \sigma\mathbf{E} \tag{2.29}$$

with D being the diffusion coefficient. The above formula gives an alternative expression for the conductivity

$$\sigma = e^2 D\frac{dn}{dE_F} \tag{2.30}$$

which is known as the Einstein relation for degenerate conductors. One can see that, the expression for conductivity here, at a first glance is significantly different from the one derived in Sect. 2. It indeed represents the alternative point of view of conductivity which is in form of a diffusion coefficient rather than mobility. In order to make sure that there is not an inconsistency between the two viewpoints, one relate the diffusion coefficient to mobility by writing it as

$$D = \frac{1}{d}v_F^2\tau \tag{2.31}$$

where d is the dimensionality of the system being considered. The overall message from this section is that, at low temperatures, electrical conduction within a metal, in the presence of significant scattering processes, being a diffusion process at the continuum level, when electrons are considered individually could be considered as a random walk process and this idea will become very important when we are discussing the concept of localization [6].

2.5 Hall Effect in a Two-Dimensional System in the Presence of Non-quantising Magnetic Field

When considering electron dynamics within a crystal and deriving Ohms law from it we have so far considered the case of zero magnetic field. In the presence of a finite but small magnetic field, because of the Lorentz force, Eq. (2.5) becomes modified as

$$m_e \frac{dv}{dt} = -eE - ev \times \mathbf{B} - m_e \frac{v}{\tau} \tag{2.32}$$

where **B** is the applied magnetic field [1, 2]. In the presence of a finite electric field in x direction, and the magnetic field in z direction, the equation yields the steady state solutions

$$\begin{aligned} m_e v_x/\tau &= -eE_x \\ 0 &= -e(E_y - v_x B) \end{aligned} \tag{2.33}$$

assuming all current remains the flow in x-direction, (i.e. $v_y = 0$). The second part of Eq. (2.37) tells us that a transverse electric field is now present in our system (depicted in Fig. 2.2) which counteracts on the Lorentz force and is therefore directly proportional it. This present transverse electric field was discovered by Edwin H. Hall in 1879 and is known as the Hall effect [7]. The proportionality constant between the electric field generated and applied magnetic field allows us to calculate a very useful quantity called the Hall coefficient. From Eq. (2.37), by substituting $j_x = nev_x$ one will obtain the Hall coefficient as

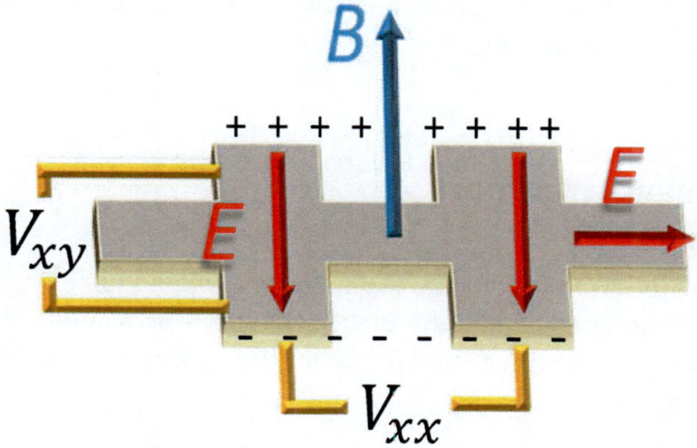

Fig. 2.2 Hall effect in the presence of a magnetic field illustrated on the most optimal device geometry to measure it, hence the name Hall bar

$$R_H = \frac{E_y}{j_x B} = -\frac{1}{ne} \tag{2.34}$$

the sign of the Hall coefficient as implied by the equation above depends on the type of dominating carriers (negative for electrons, positive for holes) hence its experimental determination allows one to find out the dominating carrier type. At this point it is very useful to introduce two very useful quantities which are longitudinal and transverse resistivity coefficients

$$E_x = \rho_{xx} J_x, \quad E_y = \rho_{xy} J_x \tag{2.35}$$

which give rise to the equations

$$\rho_{xx} = \frac{V_{xx}}{I_{SD}} \frac{L}{Wd}, \quad \rho_{xy} = \frac{V_{xy}}{I_{SD}} = -\frac{B}{ne} \tag{2.36}$$

where V_{xx} and V_{xy} are measured longitudinal and transverse voltages as depicted on Fig. 2.2, whereas I_{SD} is the source-drain current. It also has to be noted that L is the sample length at across which the voltage V_{xx} is measured, and W is the sample width across which V_{xy} is measured with d being the thickness of the system (which is absent when considering truly two-dimensional systems). An optimal device structure for Hall effect measurements is depicted in Fig. 2.2 and its known as a Hall bar. It can be seen from Eq. (2.40) that when a single carrier type is present in the system, the transverse (Hall) resistivity should vary linearly with respect to applied magnetic field. Therefore any deviations from linearity (away from quantum Hall limit which will be explained in Sect. 2.7) in transverse resistivity are either attributed anomalous Hall effect in the case of magnetic materials with single type of charge carriers or to presence of multiple charge types which can often be the case as a result of conduction across multiple bands.

2.6 Multi-band Transport

In the above section we introduced two experimentally obtainable quantities, which are namely the longitudinal and transverse resistivity, which could be defined within a tensor describing a two-dimensional system (like surface states of rhombohedral graphite thin films) as

$$\rho = \begin{pmatrix} \rho_{xx} & \rho_{xy} \\ -\rho_{xy} & \rho_{xx} \end{pmatrix} \tag{2.37}$$

We know that, by definition, $\sigma = \rho^{-1}$ and hence we obtain a conductivity tensor by inverting the resistivity tensor as

$$\sigma = \rho^{-1} = \frac{1}{\det \rho} \text{adj } \rho = \frac{1}{\rho_{xx}^2 + \rho_{xy}^2} \begin{pmatrix} \rho_{xx} & -\rho_{xy} \\ \rho_{xy} & \rho_{xx} \end{pmatrix} = \begin{pmatrix} \sigma_{xx} & \sigma_{xy} \\ -\sigma_{xy} & \sigma_{xx} \end{pmatrix}$$

where the longitudinal and transverse conductivity (σ_{xx} and σ_{xy}) are related to experimentally obtainable quantities as

$$\sigma_{xx} = \frac{\rho_{xx}}{\rho_{xx}^2 + \rho_{xy}^2} \quad \text{and} \quad \sigma_{xy} = \frac{-\rho_{xy}}{\rho_{xx}^2 + \rho_{xy}^2} \tag{2.38}$$

which after substituting expressions $\rho_{xx} = (ne\mu)^{-1}$ and $\rho_{xy} = B/ne$ become

$$\sigma_{xx} = \sum_i^N \frac{n_i e \mu_i}{1 + \mu_i^2 B^2} \quad \text{and} \quad \sigma_{xy} = \sum_i^N \frac{n_i e \mu_i^2 B}{1 + \mu_i^2 B^2} \tag{2.39}$$

for N number of carriers from different bands. Carrying out a least-squares fitting of the expressions in (2.39) to the experimental obtained magnetic field dependence of σ_{xx} and σ_{xy} until the fit converges yields one the carrier densities and mobilities of different carrier types within the system being studied.

2.7 Hall Effect in a Two-Dimensional System in the Presence of Quantising Magnetic Field

Returning to the expression for conductivity we have derived, $\sigma_0 = \frac{e^2\tau}{m_e}$, within the tensor form in the presence of magnetic field it becomes

$$\sigma = \frac{\sigma_0}{1 + \omega_C^2 \tau} \begin{pmatrix} 1 & \omega_C \tau \\ -\omega_C \tau & 1 \end{pmatrix} \tag{2.40}$$

with $\omega_C = \frac{eB}{m_e}$. Inverting the tensor above, through a similar process to what we did we get the resistivity tensor

$$\rho = \begin{pmatrix} \rho_0 & -\frac{B}{ne} \\ \frac{B}{ne} & \rho_0 \end{pmatrix} \tag{2.41}$$

where there are implications of the directly, experimentally obtainable tensor components ρ_{xy} and $-\rho_{xy}$ being independent of the time constant τ as this means that the carrier density of the system being measured could be obtained independent from scattering effects on the charges, i.e., independent of disorder of the system. In the presence of a large magnetic field, when the quantity $\omega_C \tau \sim 1$, the system enters to the semiclassical limit, when Landau levels form, because of the cyclotron motion of electrons. In this limit, the band structure of the two-dimensional system become discretised where energy levels form which are dubbed as the Landau levels. There is finite number of Landau levels that are accessible by tuning the Fermi level of the system through the gaps the system possesses. Each Landau level is indexed by a number n, and is described by the equation

$$\varepsilon_n = (n + 1/2)\hbar\omega_C = (n + 1/2)\frac{\hbar^2}{m_e l^2} \tag{2.42}$$

where the quantity l is named the magnetic length, $l = \sqrt{\frac{\hbar}{eB}}$ which was introduced in Chap. 1 when defining the harmonic oscillator after the introduction of Landau gauge as a result of finite magnetic field. The magnetic length is associated with the characteristic carrier density each Landau level can hold, which is shown below

$$n_C = \frac{1}{4\pi l^2} = \frac{eB}{h} \tag{2.43}$$

per spin per valley [8]. As each Landau level is being filled (e.g. when carrier density is tuned by a gate) or new Landau levels are created as the magnetic field is tuned one will find that the measured resistivity of the material will start oscillating, which are namely the **Shubnikov-de Haas** oscillations [9]. At a constant carrier density as the magnetic field is varied, one will see that the period of oscillations of resistivity is constant and is related to carrier density as

$$\Delta\left(\frac{1}{B}\right) = \frac{1}{n}\frac{4e}{h} \tag{2.44}$$

where the factor 4 takes care of spin and valley degeneracy for the surface states of rhombohedral graphite. Furthermore, one can obtain useful information about the system being investigated from Shubnikov de Haas (SdH) oscillations. As any other damped oscillatory wave, it can be described with a cosine function with two additional terms that take care of damping due to finite conduction lifetime of charges and finite temperature [10]

$$\frac{\Delta\rho_{xx}}{\rho_{xx}(0)} = 4\cos\left(2\pi\left(\frac{E_F}{\hbar\omega_C} - \frac{1}{2}\right)\right)\exp\left(\frac{-\pi}{\tau_q\omega_C}\right)\frac{\chi}{\sinh(\chi)} \tag{2.45}$$

where m_e is effective mass, τ_q is quantum lifetime and term χ is given by $\chi = \frac{2\pi^2 k_B T}{\hbar\omega_C}$ with T being the temperature. Quantum lifetime τ_q measures the mean time that

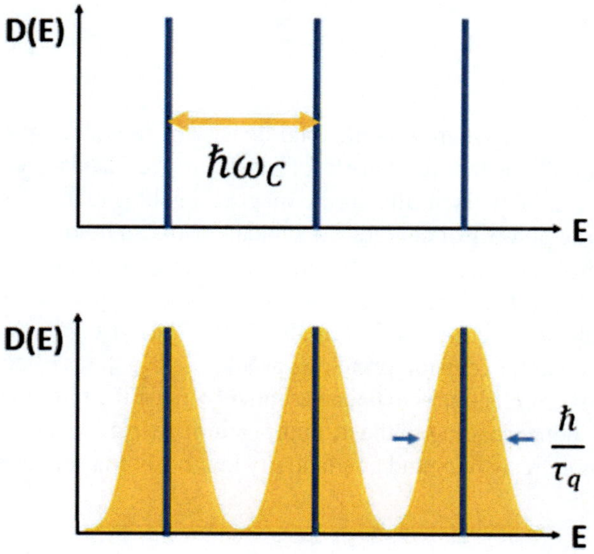

Fig. 2.3 **a** Theoretical density of states of a two-dimensional (2D) system in the presence of a Landau quantising magnetic field, with energy gaps forming because of cyclotron motion of particles. **b** Density of states of a 2D system in practice with broadened Landau levels due to scattering events within time τ_q

a carrier remains in an eigenstate before being scattered into a different state. Its main effect is to lead to a broadening of the Landau levels, which can be expressed mathematically as $\Gamma = \hbar/\tau_q$ and is depicted schematically in Fig. 2.3. By equating the expression for Landau level broadening with the expression for cyclotron energy of a given Landau level, one can derive the condition stated for the observation of SdH oscillations in the semiclassical limit, $\omega_C \tau_q \sim 1$ where it must be restated here that the scattering time in this condition is the quantum scattering time, and mobility obtained from it is named as the quantum mobility.

Quantum scattering time τ_q is sensitive to all scattering events whereas transport lifetime τ is weighed toward large angle scattering events that cause a drastic change in momentum. Ratio of quantum lifetime τ_q to transport lifetime τ (often referred to as the Dingle ratio) is used to comment on the scattering mechanisms within the system being investigated. The temperature dependent $\frac{\chi}{\sinh(\chi)}$ damping term is also very important, being first derived in a seminal paper by Lifshiftz and Kosevich [11] it is used on its own to extract the effective mass of charges from cyclotron motion by studying temperature dependence of change in longitudinal resistivity at a given magnetic field.

As magnetic field is increased, to generate Landau levels that can hold a large density of electrons, eventually all charges will occupy a given Landau level (while the Fermi level lies in a gap) with a given degeneracy aside from those of spin and valley, which is namely the filling factor [9]

$$v = \frac{n}{n_C} \tag{2.46}$$

with v being an integer where n is the total density of free electrons in the system and n_c being the characteristic density of electrons the Landau level (at a given magnetic field) can hold. Typically, as the magnetic field is increased to obtain new Landau levels, the number of charges the Landau level can hold also increases, thus leading to lower filling factors at higher magnetic fields. When the condition of the Fermi level of the system being in the gap is satisfied, the quantum scattering time τ_q is greatly enhanced, in fact, it is infinite. This is because, elastic collisions are forbidden by the Pauli exclusion principle, as all possible states of equal energy are occupied. Inelastic collisions also become unlikely given the rare presence phonons as $\hbar\omega_C \gg k_B T$ hence leading to the infinite scattering time. A natural consequence of this is vanishing resistivity and conductivity and furthermore, quantisation of the Hall resistivity as

$$\rho_{xy} = -\frac{h}{ve^2} \tag{2.47}$$

which is known as the quantum Hall effect. Experimental manifestation of quantum Hall effect, first observed by von Klitzing in 1980 [12], with both quantised transverse magnetoresistance and vanishing longitudinal resistance is shown in Fig. 2.4.

2.8 Hall Effects in the Absence of Magnetic Field

We have in Sects. 2.2 and 2.3 we have established the fact that when electrons are under the influence of a constant electric field they collectively form a steady state as a result of scattering progresses where they can be considered to have a constant drift velocity within the framework of the Drude model. Following on from there, in Sect. 2.6 we have introduced a magnetic field and the associated Lorentz force which in a nutshell gives rise to the Hall effect. Depending on the symmetry properties of the material, in the absence of either time-reversal and/or inversion symmetry it can acquire a momentum space dependent pseudo-magnetic field. This gives rise to component of group velocity of the electron wave-packets in transverse direction, as if a magnetic field associated Hall effect was present, which can be expressed in a semiclassical equation as [14]

$$v(\mathbf{k}) = \frac{1}{\hbar}\frac{\partial \varepsilon(\mathbf{k})}{\partial \mathbf{k}} + \dot{\mathbf{k}} \times \Omega(\mathbf{k}) \tag{2.48}$$

where $\Omega(\boldsymbol{k})$ is the Berry curvature density and $\dot{\mathbf{k}}$ is the rate of change of crystal momentum arising from presence of finite electric field. The fact that the electric

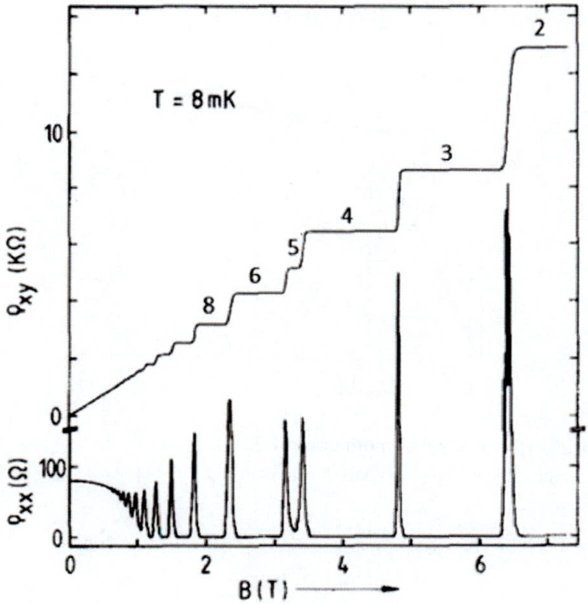

Fig. 2.4 Manifestation of quantum Hall effect on a high mobility two-dimensional electron gas system with dips in longitudinal resistivity at above 1 T and accompanied by it a quantisation of Hall resistance. Figure adapted from ref. [13]

field is applied in *x*-*y* plane, means that only *z* component of the Berry curvature density $\Omega(\boldsymbol{k})_z$ is relevant. The nature of interaction that gives rise to finite Berry curvature, also determines its properties giving rise to the specific type of intrinsic Hall effect. Berry curvature can only be present within a system, when either of the time-reversal, $\Omega_{n,k} = -\Omega_{n,-k}$ or inversion $\Omega_{n,k} = \Omega_{n,-k}$ symmetries are broken for an n-th band, however if both of the symmetries are present at the same time the Berry curvature vanishes leading to the usual equation for group velocity of an n-th band.

2.8.1 Anomalous Hall Effect

The first of the zero magnetic field Hall effects considered will be the anomalous Hall effect which was discovered by Edwin Hall in 1881 shortly after the discovery of Hall effect [15]. His observation was a larger 'pressing electricity' in magnetic iron compared to non-magnetic systems. Later, dubbed anomalous Hall effect, this phenomenon was understood to arise because of non-zero contribution to transverse resistivity arising from magnetisation M as well as the ordinary Hall effect. It is typically expressed as

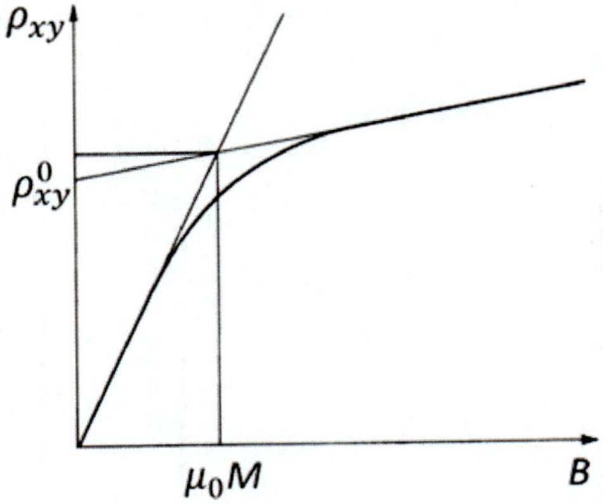

Fig. 2.5 An empirical illustration of anomalous Hall effect. A magnetisation related additional Hall resistance ρ^0_{xy} superimposed to Hall resistance yielding two separate regions with linear slopes as magnetic field is tuned

$$\rho_{xy} = R_0\mathbf{B} + \rho^0_{xy} \tag{2.49}$$

where

$$\rho^0_{xy} = R_S\mu_0\mathbf{M} \tag{2.50}$$

with **B** being the applied magnetic field and **M** being the magnetisation of the system [16]. Note that two Hall coefficients have been introduced where R_0 is known as the ordinary Hall coefficient whereas R_s is the spontaneous Hall coefficient. Experimental manifestation of anomalous Hall effect is depicted in Fig. 2.5. Two distinct linear Hall resistivity regimes are present implied by Eq. (2.49). Until a critical magnetic field (related to the magnetisation of the material) is reached, the ordinary Hall resistance of the material has the anomalous contribution superimposed.

Despite the experimental studies and the empirical relationship obtained [shown in Eq. (2.49)], the first theory which explained anomalous Hall effect intrinsically (in other words free from disorder and impurities) was put forward by Karplus and Luttinger (KL) in 1954 [17]. KL theory explained anomalous Hall effect through an anomalous velocity arising due to time-reversal symmetry breaking, which is communicated to the electrons of the system in form of spin-orbit coupling which in more modern terms became known as the Berry curvature after introduction of Berry phase to solid state physics [18].

The implications of a time-reversal symmetry broken Berry curvature i.e. $\Omega_{n,k} = \Omega_{n,-k}$, can be seen through another consideration of Eq. (2.18). The term $v(\mathbf{k})f_0$ no longer vanishes as an integral is carried out across the entire momentum space, hence leading to a transverse component of current density, which can be expressed for a single band as

$$J_H = \frac{2e^2}{\hbar}\mathbf{E} \times \int \frac{d^3k}{(2\pi)^3}\Omega(\mathbf{k}) f_0(\mathbf{k}) \tag{2.51}$$

which by Ohms Law leads to presence of a finite Hall conductivity in the presence of zero magnetic field which is expressed below [19]

$$\sigma^0_{xy} = \frac{2e^2}{h}\int \frac{d^3\mathbf{k}}{(2\pi)^3}\Omega_z(\mathbf{k}) f_0(\mathbf{k}) \tag{2.52}$$

where $\Omega_z(\boldsymbol{k})$ is the Berry curvature density acting out of plane of the sample. The most remarkable feature of the intrinsic anomalous Hall current and the conductivity obtained is that they are both independent of scattering time, hence, one can say that they are dissipation-less.

In transport experiments however, it is the Hall resistivity that is measured rather than the Hall conductivity. Knowing that the relationship between Hall resistivity and conductivity is, $\sigma'_{xy} = \frac{-\rho'_{xy}}{\rho^2_{xx}+\rho'^2_{xy}}$, and assuming that $\rho_{xx} \gg \rho'_{xy}$ one will obtain the relationship that has been found to be an experimental guide with Hall resistivity and the longitudinal resistivity being coupled to each other $\rho_{xy} \sim \rho^2_{xx}$ as the resistivity of the sample is varied through either doping or gating. Indeed this was found to be the case in the experiments observing intrinsic anomalous Hall effect [20].

On top of the intrinsic mechanism for the anomalous Hall effect, defect and impurity related extrinsic effects were also considered. The first one of these, is the skew scattering mechanism proposed by Smit in 1955 [21]. This scattering mechanism explained anomalous Hall effect to arise due to an asymmetric scattering of electrons by impurities caused by spin orbit coupling, interacting with the polarized spins of electrons as a result of iterant magnetisation of the material, hence leading to a transverse charge separation. The anomalous Hall conductivity obtained as a result of the proposed scattering mechanism was found to be proportional to the scattering time and hence a linear dependence of anomalous Hall conductivity to resistivity, $\rho_{xy} \sim \rho_{xx}$ was predicted [17]. The other scattering mechanism proposed was the side-jump contribution to the anomalous Hall conductivity. It could be regarded as an extension of the anomalous velocity theory proposed by KL to the spherical impurities with spin orbit interaction. Overall side-jump contribution was found to yield an anomalous Hall conductivity independent of scattering time. Thus, the separation between intrinsic and side jump contributions to anomalous Hall effect was for a long time the most argued aspect of anomalous Hall effect [17].

A quantised version of anomalous Hall effect was predicted by Haldane in 1988. This prediction followed the discovery of quantum Hall effect in 1980 by von Klitzing et al. [13] for which the topological nature was demonstrated by Thouless et al. [22] in 1982. Haldane, basing his model on then hypothetical material graphene, showed that in 2D systems lacking time reversal symmetry (e.g. due to spontaneous magnetic ordering) a quantized Hall conductance could be observed in the absence of any magnetic field, as a result of presence of a spontaneous Berry curvature [23]. Albeit being predicted for 2D materials which were hypothetical in 1988, experimental realisation of quantum anomalous Hall effect has so far been limited traditional thin films of a not so traditional material, namely a topological insulator, magnetically doped $Bi_2Sb_2Te_3$ [24]. Quantisation of Hall signal in absence of magnetic field was reported in this system in 2013.

2.8.2 *Spin Hall Effect*

Spin Hall effect, as the name implies means the presence of a transverse conductivity, or Hall resistivity, for spins rather than charges. This effect was first considered, unlike anomalous Hall effect, through an extrinsic mechanism by Dyakanov and Perel [25] in 1971. However, the name spin Hall effect was coined by Hirsh, who re-predicted the effect in 1999 [26]. Being closely related to anomalous Hall effect, it is theorised to occur as a result of spin-orbit interaction within both extrinsic and intrinsic considerations [26, 27].

The extrinsic contributions giving rise to the spin Hall effect are the skew and side-jump scattering mechanism as in case of extrinsic contributions to anomalous Hall conductivity, however of course, in the presence of time-reversal symmetry as no inherent magnetisation is present for the spin Hall effect. In the case of presence of a non- spin polarized current, electrons of opposite spins undergo scattering in opposite directions as a result of their inherent spin-orbit coupling property when passing by a spin-less scattering object (such as a defect). The electrons in their rest frame feel the presence of a current which processes an effective magnetic field interacting with their spin, leading to a Zeeman energy dependent force acting in counter directions for each spin. As mentioned above, this type of scattering would yield transverse spin currents that is influenced by the scattering time in the Drude model, hence it would not be dissipation-less. The side jump contribution as in the case of anomalous Hall effect, arises because of a sideways displacement of an electron wave packet due to the potential by a 'scattering' object. This mechanism as highlighted above is independent of scattering time and hence leads to dissipation-less spin currents.

The intrinsic spin Hall effect, again similarly to the intrinsic anomalous Hall effect arises in the presence of finite Berry curvature, but unlike anomalous Hall effect where the Berry curvature arises from absence of time-reversal symmetry, in the case of intrinsic spin Hall effect the Berry curvature arises from the absence of inversion symmetry which introduces a Rashba spin-orbit coupling which is a different type of spin-orbit coupling to case considered for the anomalous Hall effect [27].

The first determination of spin Hall effect both extrinsic and intrinsic nature came through experiments utilizing optical techniques [28, 29]. The way that transverse spin currents arise is naturally accompanied by charge accumulation of the same magnitude on either side of the sample in a non-spin polarized system, hence various alternative methods for electronic transport experiments were proposed [27]. Finally electronic observation happened in an extremely narrow width aluminium Hall cross with ferromagnetic contacts injecting spin polarized charges through a dielectric oxide barrier [30]. (It must be stated here that by spin polarized charges or electrical spin injection we mean creation of a non-equilibrium population of up and down spins within the system). The way Valenzuela and Tinkham determined the spin Hall effect (SHE) was through an introduction of a non-local measurement set up utilizing a Hall cross and a ferromagnetic contact. They have driven the spin polarized current away from the Hall cross which resulted in spin-polarized electrons moving towards the Hall cross. The spin Hall signal was inferred from the presence of a non-local signal (which was at the time dubbed the spin Hall voltage) at the Hall cross, as a result of charges of opposite spins moving in opposite directions towards the two edges of the sample.

Following the observation of SHE in electronic transport experiments it has been proposed that extrinsic SHE could in principle be found on paramagnetic systems with large σ_{SH}/σ_{xx}, spin conductivity to charge conductivity ratio, which is referred to as the spin Hall angle. Abanin et al. [31] have found that on systems with a much smaller charge diffusion length but much larger spin diffusion compared to the width, i.e. $l_e \ll W \ll l_s$, where W is the width of the device a non-local current will be observed without the need for spin injection through ferromagnetic contacts. They have found that this non-local signal will should the equation

$$R_{NL} = \frac{1}{2}\left(\sigma_{xy}^{S}\right)^2 \rho_{xx}^3 \frac{W}{l_S} e^{-L/l_S} \tag{2.53}$$

where L is the length of the device, σ_{xy}^{S} is the spin Hall conductivity, and l_s is the spin diffusion length. Thus, the fitting of the expression above would allow one to extract parameters l_s and spin Hall angle for the spin Hall system being investigated. The above relationship was found to be obeyed in graphene systems with enhanced spin orbit coupling, through various techniques such as hydrogenation [32]. At charge neutrality point in hydrogenated samples, values of spin diffusion length and spin Hall angle were found to be $(0.95 \pm 0.02)\mu$m and ~0.58 respectively.

A quantised version of spin Hall effect, namely the quantum spin Hall effect has been predicted theoretically and observed experimentally [33, 34]. Inspiration for the theoretical prediction of quantum spin Hall insulator came following the isolation of graphene and demonstration of its topological properties. In 2005, Kane and Mele found that inclusion of a gap, induced by spin orbit coupling, leads to a gapped phase preserving of both time-reversal and inversion symmetry as the spin dependent term describing the interaction is even under parity (inversion) and odd under time reversal. Characterised by quantised spin filtered currents, and mathematically Z2 index which has foreshadowed the prediction and experimental realisation of topological insulators, quantum spin Hall state, first observed in HgTe quantum wells has been found to give a quantised 4 terminal resistance [34]. It has to be mentioned at this point that the predicted quantum spin Hall phase in flat bands of rhombohedral graphite which arises due to electron-electron interactions, is different to the outlined well known phase which arises as a result of spin-orbit coupling interaction.

The QSH phase in these flat bands is characterised by Chern numbers rather than a Z2 index [35].

2.8.3 Valley Hall Effect

Valley Hall effect, as the name implies is the manifestation of valley Hall currents due presence of a finite Berry curvature arising from the lack of inversion symmetry in systems possessing valley degree of freedom [36]. Thus, it can be regarded as an intrinsic effect only. It could be understood to be similar to spin Hall effect through an analogy between valley and spin degree of freedom which was first introduced in traditional quantum well systems, namely AlAs grown on a GaAs substrate in 2006 [37, 38]. A valley filter based on quantum point contacts was proposed, which led to breaking of valley degeneracy in these systems.

Following this, a valley filter based on graphene (which possesses inequivalent K and K′ valleys as highlighted in Chap. 1) nanoribbons was proposed by Rycerz et al., where the term valleytronics was coined in 2007 [39]. The first theoretical prediction of (quantum) valley Hall effect came through in the same year, where in a seminal paper by Xiao, Yao and Niu the effect of inversion (sublattice) symmetry breaking was considered and was found to lead to quantised valley Hall currents [36]. This prediction was inspired by weak localization experiments showing suppression of intervalley scattering [40] and demonstration of opening of a gap in graphene due to a substrate potential [41].

As mentioned above, valley Hall effect is an intrinsic effect, and its due to absence of inversion symmetry where Berry curvature is an odd function of k, having the values of $\Omega(\boldsymbol{k})$ and $-\Omega(-\boldsymbol{k})$ at K and K′ valleys respectively, hence it can be seen by Eq. (2.51) that it leads to deflection of charges of the opposite valleys on opposite sides of the sample. The first ever experimental detection of valley Hall effect occurred in both optics and transport experiments in single layer MoS2 transistors [42] (where inversion symmetry is intrinsically absent) and graphene/hBN superlattice systems [43] (where the inversion symmetry is broken due to hBN substrate potential) in 2014. In transport experiments valley Hall effect has been analysed using a modified version of the Eq. (2.53) which is

$$R_{NL} = \frac{1}{2}\left(\sigma_{xy}^{V}\right)^2 \rho_{xx}^3 \frac{W}{l_V} e^{-L/l_V} \tag{2.54}$$

where σ_{xy}^{v} is the valley Hall effectivity and $\alpha = \sigma_{xy}^{v}\rho_{xx}$ is the valley Hall angle, analogous with the spin Hall angle. The above model has shown to be valid for the case $\alpha 1$ in reference [43] and has been analytically obtained for the cases $\alpha \ll 1$ and $\alpha^2 \ll 1$ in references [44, 45] respectively. Overall, it is now generally accepted that in a system honeycomb lacking inversion symmetry, a cubic dependence of non-local signal to resistivity corresponds to valley Hall effect and associated Berry curvature [43–47]. Quantum valley Hall effect is expected arise in the absence of intervalley scattering, where the Hall conductivity as a result of valley Hall currents is expected to approach a quantized value, which has been observed on 1D AB-BA domain walls on bilayer graphene systems where the two terminal conductance of the devices fabricated have been found to approach a quantised value for smaller sized domain walls [48].

An interesting property which also emerges from inversion asymmetry in honeycomb systems is orbital magnetic moment and associated orbital magnetisation, which have opposite signs in opposing valleys. The orbital magnetic moment associated with a gapped two band system has been shown to be [36]

$$m_z(\mathbf{p}, \tau_z, s_z) = -\tau_z \frac{\Delta}{\sqrt{\Delta^2 + E(\mathbf{p})^2}} \left(\frac{\partial}{\partial \mathbf{p}} E(\mathbf{p}) \right)^2 m_e \tag{2.55}$$

where $\tau_z = \pm 1$ is the valley index and $s_z = \pm 1$ is the spin index and Δ is the gap which may be both spin and valley dependent. Such an orbital magnetic moment, and resulting magnetisation of opposite values in opposite valleys has been probed in optics experiments [49–52].

2.9 Localization Effects in Disordered Mesoscopic Systems

The idea of diffusive process of electron transport is central to understanding the phenomena of weak localization (happening in weak disorder limit). Hence, we will first review the classical theory of diffusive process and its dimensionality to introduce the concept of localization and will later consider quantum effects in weak disorder limit, highlighting weak (anti)localization in the presence of disorder in mesoscopic systems.

2.9.1 Classical Picture of Diffusion and Localization

Let's consider the classical diffusion of an electron in a disordered d dimensional crystal system. The random walk of the electron through the system takes place as it is elastically scattered off various impurities. We say that such an electron is localized if it has non-zero probability of returning to its original position (at $t = 0$) as $t -> \infty$ [53].

For a probability distribution function of a system possessing such a particle, $p(\mathbf{r}, t)$, the diffusion equation could be written as

$$\frac{\partial p}{\partial t} - D\nabla_{\mathbf{r}} p = 0 \tag{2.56}$$

which leads to the solution

$$p(\mathbf{r}, t) = (4\pi Dt)^{-d/2} \exp\left(\frac{-\mathbf{r}^2}{4\pi Dt}\right) \tag{2.57}$$

which means that the probability to return to origin at a time t in a d dimensional system is

$$p(0, t) = (4\pi Dt)^{-d/2} \tag{2.58}$$

where the integrated probability, for a particle to return to origin after a first scattering event, at a time τ being

$$P(0, t) = \left(\frac{1}{4\pi D}\right)^{d/2} \int_{\tau}^{t} \frac{dt}{t^{d/2}} \propto \begin{cases} (t/\tau)^{1/2} & d = 1 \\ \ln(t/\tau) & d = 2 \\ (t/\tau)^{-1/2} & d = 3 \end{cases} \tag{2.59}$$

from which and one can immediately see that as t approaches infinity, localization is evident in 1D and 2D cases, whereas for the 3D case the electrons remain delocalized. Once one introduces quantum mechanics, the classical picture highlighted above becomes modified, and one obtains an enhanced back-scattering probability in a weakly disordered system which is known as the weak localization.

2.9.2 *Weak Localization*

We have so far, when considering their dynamics within a crystal in zero or low magnetic fields, assumed electrons as classical particles, especially within the Drude model. It was the seminal work of Anderson in 1958 [54] that showed the fact that significant deviations from classical picture may occur due quantum nature of electrons in a strongly disordered crystal lattice, leading to in fact, a complete absence of diffusion in these lattices, manifesting itself as a metal-insulator transition, which became known as Anderson or strong localization [54]. A few decades later, in 1980, considering quantum dynamics of electrons in a weakly disordered metallic system (which is where the term weak comes from), two groups have predicted a distinct temperature and magnetic field dependence of conductivity at low temperatures in two dimensional systems [55, 56], predicting weak (anti)localization in a metallic system which was also extended to three dimensions shortly after [57].

When considering quantum dynamics of electrons, we are dealing with probability amplitudes, which have wavelengths and frequencies that relate to momenta and energies, respectively. These momenta and energies, rather than forces, are incorporated in a quantum system through phases of the wavefunctions. The most central concepts

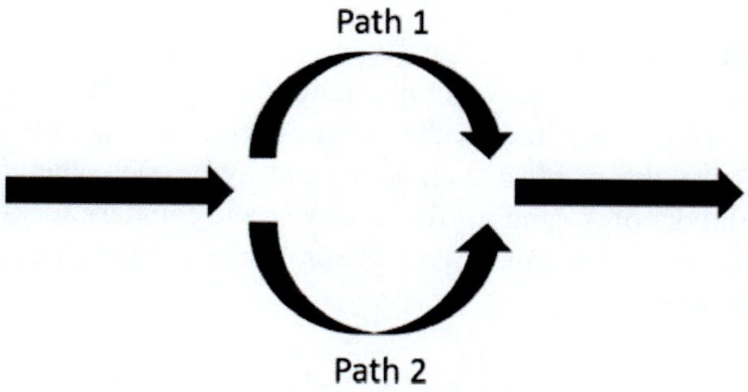

Fig. 2.6 Two electron beams splitting to take alternative circular paths before merging into another single beam

to understanding weak localization are the terms dephasing time and the dephasing length which are related to the wavelike nature of electrons. The simplest way of illustrating the concepts of dephasing time and dephasing length is through an imaginary interference experiment of two electron beams as illustrated in Fig. 2.6. In the absence of magnetic field, each electron, undertaking path 1 and 2 respectively, could be described by plane waves, $A_a = |A_a|e^{i\varphi_a}$ and $A_b = |A_b|e^{i\varphi_b}$. Now, if we consider a static system, in which electrons travelling through each pathway undergo scattering due to collisions with impurities the phase relationship at merging point X will always be the same. At low enough temperatures, it has been found that the dominant interaction effect that influences the phases of such electrons with different trajectories are inelastic scattering events such as the electron-electron interactions and electron-phonon interactions. Remarkably, the timescale at which such an inelastic scattering event occurs, which is effective at destroying an electrons phase, τ_φ is much larger than the transport relaxation time τ. Thus, electrons undergoing many scattering events within a crystal do not lose their phase information until an inelastic scattering event, which happens much less frequently. Hence, an obvious approach at defining a dephasing length by writing the equation

$$L_\varphi = v_F \tau_\varphi \tag{2.60}$$

is only true for very high mobility systems where $\tau_\varphi \sim \tau$. In low mobility systems, where $\tau_\varphi \gg \tau$, as velocity is randomised many times before a dephasing inelastic scattering event occurs, one needs to consider the root mean squared distance, which for a two dimensional system is given by [6]

$$L_\varphi^2 = \frac{\tau_\varphi}{\tau}(v_F \tau)^2 \langle \cos^2(\theta) \rangle = \frac{v_F^2 \tau \tau_\varphi}{2} = D\tau_\varphi \tag{2.61}$$

where θ is the scattering angle at each collision, which leads to the following

$$L_\varphi = \sqrt{D\tau_\varphi} \tag{2.62}$$

as the expression for the dephasing length.

Let's imagine an electron undergoing a random walk within a crystal lattice to return to its original position and for the sake of keeping our argument simple, let's consider it doing that through two alternative pathways to each other, which have the probability amplitudes of $A_{\subset}$ and $A_{\supset}$ for clockwise and counter-clockwise directions respectively. Classically one would write the probability of the electron returning to its original position as

$$P_{11} = |A_{\subset}|^2 + |A_{\supset}|^2 \tag{2.63}$$

which would be correct only if we ignore the quantum mechanical nature of electrons. Taking care of the quantum mechanics of electrons, we would have to write the probability of electrons returning to their original position as

$$P_{11} = |A_{\subset}|^2 + |A_{\supset}|^2 + 2|A_{\subset}A_{\supset}|\cos(\varphi_{\subset} - \varphi_{\supset}) \tag{2.64}$$

where φ is the phase of the electron as it is completing the clockwise and anticlockwise pathways, respectively. Given that the random walks are carried out in opposite directions in same system, one can write the amplitudes as

$$A = A_{\subset} = A_{\supset}$$

and the phase difference term assuming time scale t is much smaller than dephasing length as

$$\cos(\varphi_{\subset} - \varphi_{\supset}) = 1$$

which would lead to a probability amplitude given as

$$P_{11} = 4|A|^2 \tag{2.65}$$

being double the probability one obtains classical relation given in Eq. (2.63).

Hence, we saw that, during diffusive conduction across a mesoscopic sample, in the presence of time reversal symmetry, the probability of the occurrence of rare electron paths (in which electrons complete closed cycles to return to their original position) is doubled to due to their interference with their counter clockwise partners. This leads to a doubled amplitude of electron scattering at a zero angle to the x-axis towards which the diffusive motion within the system is taking place as illustrated in Fig. 2.7. Remarkably, one obtains this enhancement as a small and negative correction to the Drude conductivity

$$G = G_{Drude} + \Delta G_{WL} \tag{2.66}$$

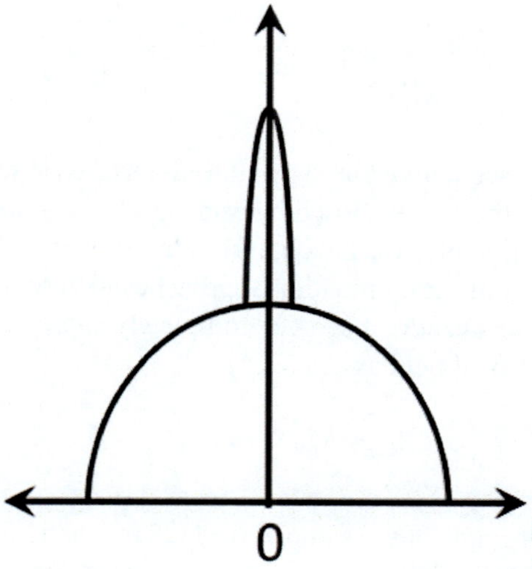

Fig. 2.7 Backward scattered wave amplitudes as a particle is undertaking a diffusive motion towards a defined *x*-axis with a doubling happening for backward scattering happening exactly at zero-degree angle

where ΔG_{WL} is on the order of $-e^2/h$.

Excitingly from an experimental point of view, the story does not end here. The weak localisation effect of enhanced backward scattering is found to be remarkably sensitive to magnetic field, or one should say rather to introduction of a vector potential to the system. The fact that the weak localization correction is destroyed by application of a magnetic field has established it as unique signature of this phenomena.

In the presence of a magnetic field, the phases of the wavefunctions describing the electrons undergoing a random walk in clockwise and anticlockwise directions become modified as

$$\varphi_{\subset} = \varphi_{\subset(0)} + \frac{e}{\hbar}\oint\limits_{\subset} \mathbf{A}\cdot d\mathbf{r} \quad \text{and} \quad \varphi_{\supset} = \varphi_{\supset(0)} + \frac{e}{\hbar}\oint\limits_{\supset} \mathbf{A}\cdot d\mathbf{r} \tag{2.67}$$

where the phase difference between two counterpropagating paths becomes

$$\varphi_{\subset} - \varphi_{\supset} = \frac{2e}{\hbar}\oint \mathbf{A}\cdot d\mathbf{r} = \frac{2e}{\hbar}\iint\limits_{S} (\nabla\times\mathbf{A})\cdot d\mathbf{S} = \frac{2e}{\hbar}BS \tag{2.68}$$

with S being the area enclosed by the path. Thus, we can write the quantum interference term in Eq. (2.67), which we shall call I, for n number of closed paths as

$$I = \sum_n 2A^2 \cos\left(\frac{2e}{\hbar} BS\right) = \sum_n 2A^2 \cos\left(\frac{2\pi}{B_F} B\right) \tag{2.69}$$

where one can see that we have derived an oscillatory interference term which has a frequency that depends on the area enclosed by each specific counter propagating loop pairs. Given that, in principle, we have an infinite number of loops all with varying loop areas, the result of the summation approaches to zero above a critical magnetic field where weak localization becomes completely supressed. One can do a rough estimation of this critical field as

$$B_C \sim h/eS \tag{2.70}$$

where S could be taken as the area of the sample to which magnetic field is pointing away from, as a rough estimation, but it is the size of the largest loop charge carriers carry out whilst undergoing a random walk.

In the most simplistic scenario following scaling relationships have been derived for the change in conductivity with the application of a magnetic field due to suppression of weak localization

$$\Delta\sigma(B) \propto \begin{cases} \frac{e^2}{h} \ln\left(\frac{eBD\tau_\varphi}{\hbar}\right) \\ \frac{e^2}{h}\sqrt{\frac{eB}{\hbar}} \end{cases} \tag{2.71}$$

in 2 and 3 dimensions, respectively [58]. It must be noted that in 3D, the suppression of the weak localisation due to magnetic field is independent of its orientation with respect to current.

2.9.3 *Weak Anti-localization*

Hand in hand with the prediction of weak localization, an interference induced positive correction to conductivity was also predicted, arising from reduction of the probability of backward scattering. This phenomenon was initially predicted to arise in materials possessing strong spin orbit coupling, where there is a spin-momentum locking [56, 59]. The effect in a nutshell arises as two wavelike particles possessing spins (locked to the direction of momentum) gain opposite phases as they are carrying out loops in counter-clockwise directions. They interfere destructively leading to the suppression (halving) of the classical probability amplitude for random walk in backward direction.

More recently, after the discovery of graphene, weak anti-localization effect was also demonstrated in this carbon based system [60, 61] which intrinsically has very weak spin-orbit coupling effect. The observation of the effect in graphene was indeed due to presence of a Berry phase, and associated coupling of pseudospin degree of freedom to momentum. Analogous to the case of spin-orbit coupling, quasiparticles carrying out counter-clockwise loops gain a Berry phase of opposite values hence leading to a destructive interference.

Over the past decade, following on the breakthrough with regards to experimental realisation of topological insulators and Weyl/Dirac semimetals, weak antilocalization has also been observed in these systems, both of surface and bulk origins in the former, and being of bulk 3D origin in the latter [62–65]. Interestingly, possessing surface bands that are N-layer generalisation of graphene, and possessing bulk subbands that are gapped bulk Dirac cones, ABC graphite films are expected to show an interesting interplay of weak (anti)-localization effects in both two and three dimensions. The requirement of sufficiently low temperatures also holds for weak anti-localization as one needs to have a phase coherence length much larger than mean free path. The characteristic experimental signature for suppression of weak anti-localization due to presence of a magnetic field has been, theoretically shown to be

$$\Delta\sigma(B) \propto \frac{e^2}{h}\sqrt{\frac{eB}{\hbar}} \tag{2.72}$$

in 3D at large magnetic fields. For 2D, the situation is more complicated and an expression involving digamma functions has been derived and used to fit experiments [56]. For graphene, this expression has been shown to be [60, 61]

$$\Delta\sigma(B) = \frac{e^2}{\pi h}\left(F\left(\frac{\hbar/4eDB}{1/\tau_\varphi}\right) - F\left(\frac{\hbar/4eDB}{\frac{1}{\tau_\varphi}+\frac{2}{\tau_i}}\right) - F\left(\frac{\hbar/4eDB}{\frac{1}{\tau_\varphi}+\frac{1}{\tau_i}+\frac{1}{\tau_*}}\right)\right) \tag{2.73}$$

where $F(z) = \ln(z) + \Psi(0.5 + z^{-1})$ with $\Psi(x)$ being a digamma function. Constants τ_* and τ_i are intravalley and intervalley scattering times, respectively. The sign of the term in brackets determines the sign of the correction to conductivity.

References

1. Hook JR, Hall HE (1990) Solid state physics. Wiley
2. Kittel C (2004) Introduction to solid state physics. Wiley
3. Bostwick A, Ohta T, Seyller T, Horn K, Rotenberg E (2007) Quasiparticle dynamics in graphene. Nat Phys 3(1):36–40. https://doi.org/10.1038/nphys477
4. Drude P (1900) Zur Elektronentheorie der Metalle. Ann Phys 306(3):566–613. https://doi.org/10.1002/andp.19003060312
5. Bloch F (1929) Über die Quantenmechanik der Elektronen in Kristallgittern. Zeitschrift für Phys 52(7–8):555–600. https://doi.org/10.1007/BF01339455
6. Datta S (1997) Electronic transport in mesoscopic systems. Cambridge University Press
7. Hall EH (1879) On a new action of the magnet on electric currents. Am J Math 2(3):287–292

8. Prange RE, Girvin SM (1987) The quantum hall effect. Springer, New York
9. Schubnikow WJ, de Haas L (1930) Neue Erscheinungen bei der Widerstandsänderung von Wismuthkristallen im Magnetfeld bei der Temperatur von flüssigem Wasserstoff (I). Proc R Netherlands Acad Arts Sci 33:363–378
10. Coleridge PT, Stoner R, Fletcher R (1989) Low-field transport coefficients in GaAs/Ga1-xAlxAs heterostructures. Phys Rev B 39(2):1120–1124. https://doi.org/10.1103/PhysRevB.39.1120
11. Kosevich AM, Lifshitz IM (1956) The de Haas-van Alphen effect in thin metal layers. Jetp 2(4):646–649
12. Klitzing KV, Dorda G, Pepper M (1980) New method for high-accuracy determination of the fine-structure constant based on quantized hall resistance. Phys. Rev. Lett. 45(6):494–497. https://doi.org/10.1103/PhysRevLett.45.494
13. Klitzing KV (1984) The quantized Hall effect. Physica B 126:242–249. https://doi.org/10.1016/0378-4363(84)90170-0
14. Chang MC, Niu Q (1995) Berry phase, hyperorbits, and the Hofstadter spectrum. Phys Rev Lett 75(7):1348–1351. https://doi.org/10.1103/PhysRevLett.75.1348
15. Hall EH (1881) On the 'rotational coefficient' in nickel and cobalt. London Edinburgh Dublin Philos Mag J Sci 12(74):157–172. https://doi.org/10.1080/14786448108627086
16. Nagaosa N, Sinova J, Onoda S, MacDonald AH, Ong NP (2010) Anomalous Hall effect. Rev Mod Phys 82(2):1539–1592. https://doi.org/10.1103/RevModPhys.82.1539
17. Karplus R, Luttinger JM (1954) Hall effect in Ferromagnetics. Phys Rev 95(5):1154–1160. https://doi.org/10.1103/PhysRev.95.1154
18. Berry MV (1984) Quantal phase factors accompanying adiabatic changes. Proc R Soc A Math Phys Eng Sci 392(1802):45–57. https://doi.org/10.1098/rspa.1984.0023
19. Jungwirth T, Niu Q, MacDonald AH (2002) Anomalous hall effect in ferromagnetic semiconductors. Phys Rev Lett 88(20):4. https://doi.org/10.1103/PhysRevLett.88.207208
20. Lee WL, Watauchi S, Miller VL, Cava RJ, Ong NP (2004) Dissipationless anomalous hall current in the ferromagnetic spinel CuCr2Se4-xBrx. Science (80-) 303(5664):1647–1649. https://doi.org/10.1126/science.1094383
21. Smit J (1955) The spontaneous hall effect in ferromagnetics I. Physica 21(6–10):877–887. https://doi.org/10.1016/S0031-8914(55)92596-9
22. Thouless DJ, Kohmoto M, Nightingale MP, Den Nijs M (1982) Quantized hall conductance in a two-dimensional periodic potential. Phys Rev Lett 49(6):405–408. https://doi.org/10.1103/PhysRevLett.49.405
23. Haldane FDM (1988) Model for a quantum hall effect without landau levels: condensed-matter realization of the 'parity anomaly.' Phys Rev Lett 61(18):2015–2018. https://doi.org/10.1103/PhysRevLett.61.2015
24. Chang C-Z et al (2013) Experimental observation of the quantum anomalous hall effect in a magnetic topological insulator. Science 340(6129):167–170. https://doi.org/10.1126/science.1234414
25. D'yakonov MI, Perel VI (1971) Possibility of orienting electron spins with current. Sov Phys JETP Lett 13(11):467. Available: http://www.jetpletters.ac.ru/ps/1587/article_24366.shtml
26. Hirsch JE (1999) Spin hall effect. Phys Rev Lett 83(9):1834–1837. https://doi.org/10.1103/PhysRevLett.83.1834
27. Sinova J, Culcer D, Niu Q, Sinitsyn NA, Jungwirth T, Macdonald AH (2004) Universal Intrinsic spin hall effect, no March, pp 1–4. https://doi.org/10.1103/PhysRevLett.92.126603
28. Kato YK, Myers RC, Gossard AC, Awschalom DD (2004) Observation of the spin hall effect in semiconductors. Science 306(5703):1910–1913. https://doi.org/10.1126/science.1105514
29. Wunderlich J, Kaestner B, Sinova J, Jungwirth T (2005) Experimental observation of the spin-hall effect in a two-dimensional spin-orbit coupled semiconductor system. Phys Rev Lett 94(4):1–4. https://doi.org/10.1103/PhysRevLett.94.047204
30. Valenzuela SO, Tinkham M (2006) Direct electronic measurement of the spin Hall effect. Nature 442(7099):176–179. https://doi.org/10.1038/nature04937

31. Abanin DA, Shytov AV, Levitov LS, Halperin BI (2009) Nonlocal charge transport mediated by spin diffusion in the spin Hall effect regime. Phys Rev B-Condens Matter Mater Phys 79(3):77–81. https://doi.org/10.1103/PhysRevB.79.035304
32. Balakrishnan J, Kok Wai Koon G, Jaiswal M, Castro Neto AH, Özyilmaz B (2013) Colossal enhancement of spin-orbit coupling in weakly hydrogenated grapheme. Nat Phys 9(5):284–287. https://doi.org/10.1038/nphys2576
33. Kane CL, Mele EJ (2005) Quantum Spin hall effect in graphene. Phys Rev Lett 95(22):1–4. https://doi.org/10.1103/PhysRevLett.95.226801
34. Konig M et al (2007) Quantum spin hall insulator state in HgTe quantum wells. Science 318(5851):766–770. https://doi.org/10.1126/science.1148047
35. Zhang F, Jung J, Fiete GA, Niu Q, MacDonald AH (2011) Spontaneous quantum hall states in chirally stacked few-layer graphene systems. Phys Rev Lett 106(15):1–4. https://doi.org/10.1103/PhysRevLett.106.156801
36. Xiao D, Yao W, Niu Q (2007) Valley-contrasting physics in graphene: magnetic moment and topological transport. Phys Rev Lett 99(23):1–4. https://doi.org/10.1103/PhysRevLett.99.236809
37. Gunawan O, Shkolnikov YP, Vakili K, Gokmen T, De Poortere EP, Shayegan M (2006) Valley susceptibility of an interacting two-dimensional electron system. Phys Rev Lett 97(18):1–4. https://doi.org/10.1103/PhysRevLett.97.186404
38. Gunawan O, Habib B, De Poortere EP, Shayegan M (2006) Quantized conductance in an AlAs two-dimensional electron system quantum point contact. Phys Rev B-Condens Matter Mater Phys 74(15):1–8. https://doi.org/10.1103/PhysRevB.74.155436
39. Rycerz A, Tworzydło J, Beenakker CWJ (2007) Valley filter and valley valve in graphene. Nat Phys 3(3):172–175. https://doi.org/10.1038/nphys547
40. Morozov SV et al (2006) Strong suppression of weak localization in graphene. Phys Rev Lett 97(1):7–10. https://doi.org/10.1103/PhysRevLett.97.016801
41. Zhou SY et al (2007) Substrate-induced bandgap opening in epitaxial graphene. Nat Mater 6(10):770–775. https://doi.org/10.1038/nmat2003
42. Mak KF, McGill KL, Park J, McEuen PL (2014) The valley hall effect in MoS2 transistors. Science 344(6191):1489–1492. https://doi.org/10.1126/science.1250140
43. Gorbachev RV et al (2014) Detecting topological currents in graphene superlattices. Science 346(6208):448–451. https://doi.org/10.1126/science.1254966
44. Shimazaki Y, Yamamoto M, Borzenets IV, Watanabe K, Taniguchi T, Tarucha S (2015) Generation and detection of pure valley current by electrically induced Berry curvature in bilayer graphene. Nat Phys 11(12):1032–1036. https://doi.org/10.1038/nphys3551
45. Hung TYT, Camsari KY, Zhang S, Upadhyaya P, Chen Z (2019) Direct observation of valley-coupled topological current in MoS2. Sci Adv 5(4):1–7. https://doi.org/10.1126/sciadv.aau6478
46. Sui M et al (2015) Gate-tunable topological valley transport in bilayer graphene. Nat Phys 11(12):1027–1031. https://doi.org/10.1038/nphys3485
47. Wu Z et al (2019) Intrinsic valley Hall transport in atomically thin MoS2. Nat Commun 10(1):1–8. https://doi.org/10.1038/s41467-019-08629-9
48. Ju L et al (2015) Topological valley transport at bilayer graphene domain walls. Nature 520(7549):650–655. https://doi.org/10.1038/nature14364
49. Aivazian G et al (2015) Magnetic control of valley pseudospin in monolayer WSe 2. Nat Phys 11(2):148–152. https://doi.org/10.1038/nphys3201
50. Srivastava A, Sidler M, Allain AV, Lembke DS, Kis A, Imamoḡlu A (2015) Valley Zeeman effect in elementary optical excitations of monolayer WSe 2. Nat Phys 11(2):141–147. https://doi.org/10.1038/nphys3203
51. Macneill D et al (2015) Breaking of valley degeneracy by magnetic field in monolayer MoSe2. Phys Rev Lett 114(3):1–5. https://doi.org/10.1103/PhysRevLett.114.037401
52. Li Y et al (2014) Valley splitting and polarization by the Zeeman effect in monolayer MoSe2. Phys Rev Lett 113(26):1–5. https://doi.org/10.1103/PhysRevLett.113.266804

53. Duan F, Guojun J (2005) Introduction to condensed matter physics: volume 1, no. v. 1. World Scientific
54. Anderson PW (1958) Absence of diffusion in certain random lattices. Phys Rev 109(5):1492–1505. https://doi.org/10.1103/PhysRev.109.1492
55. Altshuler BL, Khmel'nitzkii D, Larkin AI, Lee PA (1980) Magnetoresistance and Hall effect in a disordered two-dimensional electron gas. Phys Rev B 22(11):5142–5153. https://doi.org/10.1103/PhysRevB.22.5142
56. Hikami S, Larkin AI, Nagaoka Y (1980) Spin-orbit interaction and magnetoresistance in the two dimensional random system. Prog Theor Phys 63(2):707–710. https://doi.org/10.1143/ptp.63.707
57. Kawabata A (1980) Theory of negative magnetoresistance in three-dimensional systems. Solid State Commun 34(6):431–432. https://doi.org/10.1016/0038-1098(80)90644-4
58. Al'tshuler BL, Aronov AG, Larkin AI, Khmel'nitskii DE (1981) The anomalous magnetoresistance in semiconductors. Sov Phys JETP 54(2):0411. Available: http://www.jetp.ac.ru/cgi-bin/e/index/e/54/2/p411?a=list
59. Bergmann G (1982) Weak anti-localization—an experimental proof for the destructive interference of rotated spin. Solid State Commun 42(11):815–817. https://doi.org/10.1016/0038-1098(82)90013-8
60. Mccann E, Kechedzhi K, Fala'Ko VI, Suzuura H, Ando T, Altshuler BL (2006) Weak-localization magnetoresistance and valley symmetry in grapheme. Phys Rev Lett 97(14):14–17. https://doi.org/10.1103/PhysRevLett.97.146805
61. Tikhonenko FV, Kozikov AA, Savchenko AK, Gorbachev RV (2009) Transition between electron localization and antilocalization in graphene. Phys Rev Lett 103(22):1–4. https://doi.org/10.1103/PhysRevLett.103.226801
62. Kim HJ et al (2013) Dirac versus Weyl fermions in topological insulators: Adler-Bell-Jackiw anomaly in transport phenomena. Phys Rev Lett 111(24):1–5. https://doi.org/10.1103/PhysRevLett.111.246603
63. Huang X et al (2015) Observation of the chiral-anomaly-induced negative magnetoresistance: In 3D Weyl semimetal TaAs. Phys Rev X 5(3):1–9. https://doi.org/10.1103/PhysRevX.5.031023
64. Zhang CL et al (2016) Signatures of the Adler-Bell-Jackiw chiral anomaly in a Weyl fermion semimetal. Nat Commun 7:1–9. https://doi.org/10.1038/ncomms10735
65. Li Q et al (2016) Chiral magnetic effect in ZrTe 5. Nat Phys 12(6):550–554. https://doi.org/10.1038/nphys3648

Chapter 3
Experimental Technicalities

3.1 Fabrication of van der Waals Heterostructures

Following the isolation of graphene in 2004 [1], its electronic properties were investigated through transport devices fabricated with graphene flakes on Si/SiO_2 wafers [2, 3]. A major drawback of SiO_2 as a substrate was the limitation of sample quality (impacting parameters such as charge mobility) due to the influence of its surface charges [4], surface roughness [5, 6] and optical phonons [7]. This limitation was overcome by suspending the graphene flakes over the SiO_2 substrate [8], which however meant that there were restrictions on the device designs.

Shortly after the isolation of graphene, Novoselov et al. [9] in 2005, had shown that the mechanical exfoliation method could be extended to the other layered van der Waals crystals. Out of these crystals, hexagonal boron nitride (hBN) was found to be a wide band gap insulator [10] even in the monolayer limit, with a lattice parameter very close to the one of graphene [11]. In 2010, utilising hBN as a substrate instead of SiO_2, it was shown that the charge mobility in transport devices was improved by an order of magnitude, with the roughness of the device being reduced by a factor of 3, approaching to the atomic limit [12].

The method used in the work by Dean et al. [12], to transfer graphene on top of a hBN flake acting as a substrate was the so called 'wet transfer' method. The hBN flake to act as a substrate was exfoliated onto a Si/SiO_2 wafer. Graphene was then exfoliated onto a Si/SiO_2 wafer coated with a water-soluble layer and a PMMA membrane. Once the graphene flake was selected the water-soluble layer was dissolved in a deionised water bath, leaving the flake and the PMMA membrane floating. The floating PMMA membrane was picked up using a glass slide and transferred onto the previously identified hBN substrate, using a transfer stage consisting of an optical microscope and a micromanipulator. Once the transfer process was completed the PMMA membrane was dissolved using acetone.

To achieve a greater sample quality, the 'wet transfer' process was modified such that the graphene layer of interest was prevented from contacting with any solvent [13, 14]. In some cases, this was done through employing an alternative membrane to the

S. Ozdemir, *Electronic Properties of Rhombohedral Graphite*, Springer Theses,
https://doi.org/10.1007/978-3-030-88307-2_3

water-soluble layer that the PMMA layer used could peel off from. This approach and the device quality obtained as a result has led to observation of fractional quantum Hall effect as well as room temperature ballistic transport in graphene [13, 14]. Nevertheless, albeit the subsequent improvements made in the fabrication procedure to achieve a 'dry transfer' a wet chemical had to be employed at some stage in the fabrication process [15]. Overall, there was room for development of a time-efficient all dry fabrication process.

An all dry-process with an improved time efficiency was developed in 2014 initially utilising PDMS membrane and its viscoelastic nature, and later in 2015 with a combination of PMMA/PDMS membranes and their different thermal expansion coefficients with respect to the van der Waals crystals hBN and (multilayer) graphene [16]. The latter process (depicted in Fig. 3.1) has been found to lead to a

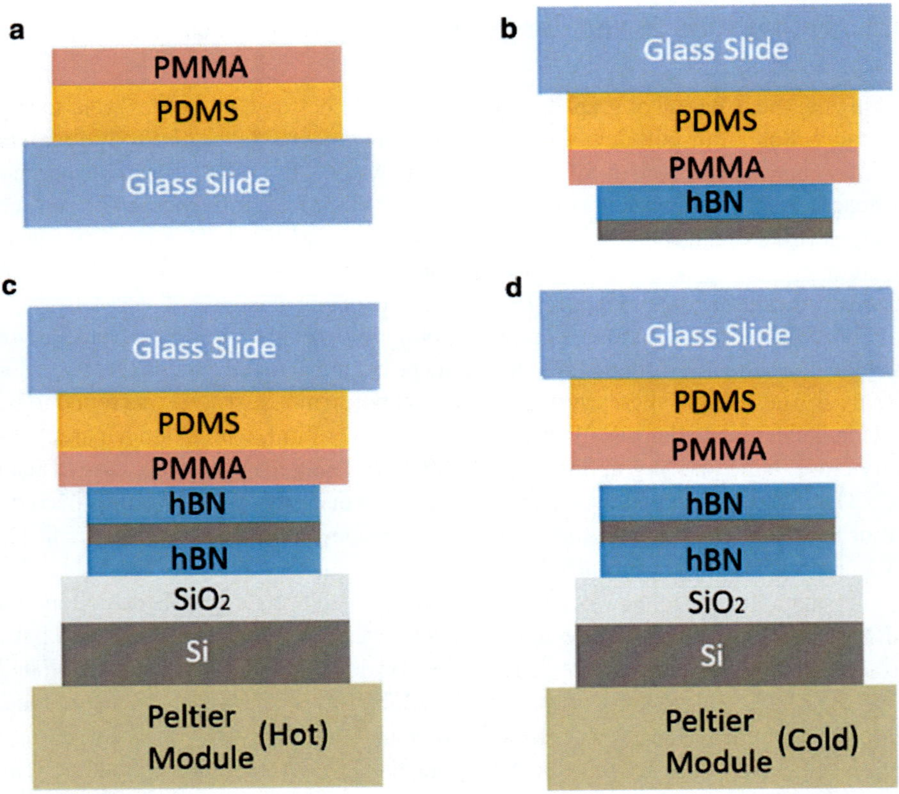

Fig. 3.1 All dry fabrication illustrated schematically. **a** Prepared PDMS/PMMA heterostructure on a glass slide onto which top hBN is mechanically exfoliated. **b** Lifting of the previously identified rhombohedral graphite flake following the mounting of the glass slide onto a transfer stage equipped with an optical microscope and a micro-manipulator. **c** Formation of the hBN/RG/hBN heterostructure after dropping the heterostructure formed in **b** to the bottom hBN flake. At this stage Peltier module is heated to ~55 °C causing thermal expansion of all layers. **d** Peltier module is cooled to ~15 °C, enabling removal of PMMA/PDMS from the top hBN leaving behind the desired heterostructure

cleaner hBN/graphene/hBN interface. A property of hBN crystals as substrates is that through their self-cleansing mechanism [17], they interact with the desired crystal forming an atomically smooth interface and hence pushing away any contaminations on the interface to regions where bubbles of these contaminations are formed. The process depicted in Fig. 3.1 was found to lead to reduced number of bubbles, hence suggesting a cleaner interface following the double encapsulation of the 2D crystals. It was used to fabricate the rhombohedral graphite heterostructures investigated in this thesis.

The fabrication procedure, depicted in Fig. 3.1, was as follows. Crystals of hBN and graphite were mechanically exfoliated onto SiO_2(290 nm)/Si wafers with their respective thicknesses being obtained through combination of optical contrast and atomic force microscopy. Rhombohedral domains of graphite films were identified through Raman mapping, in particularly FWHM of the 2D peak which couples to the electronic structure of the crystal. Following exfoliation of crystals, the top hBN layer was picked up by a prepared PDMS/PMMA heterostructure on a glass slide (depicted in Fig. 3.1a). Using the glass/PDMS/PMMA/hBN heterostructure flakes mounted on a transfer stage, rhombohedral graphite domains were picked up (as shown in Fig. 3.1b). This heterostructure was then dropped onto bottom hBN crystal exfoliated onto Si/SiO_2 substrate. At this point, using the Peltier module on which the Si/SiO_2 wafer was transferred to, the whole heterostructure was heated to ~55 °C. An obvious concern here would be potential transition of rhombohedral graphite domains into more energetically stable Bernal hexagonal phase. We have previously shown by Joule heating experiments that rhombohedral graphite domains are stable up to temperatures approaching 227 °C [18], hence the temperature of 55 °C is low enough to keep the rhombohedral domains stable. The heating of the heterostructure is done to prevent picking of the hBN flakes on the Si/SiO_2 substrate by the PMMA membrane. Following this, the Peltier module was cooled down to ~15 °C and Glass/PDMS/PMMA layers were peeled off from the top hBN flake (Fig. 3.1d). The mechanism of peeling off is attributed to the thermal shrinkage of the PDMS membrane to which PMMA is attached. Once the van der Waals heterostructure was formed, standard e-beam lithography, reactive ion etching, and metal contact deposition processes were used to form the Hall bar devices.

3.2 Measurement Electronics

When carrying out electronic measurements to characterise simple Hall bars one typically has two options which are AC and DC. It is often much more practical to apply AC currents and detect AC voltages especially working with low resistances (up to ~30 kΩ for the case of rhombohedral graphite). At low temperatures, one often needs to apply very low currents and in return detect very low voltages (can be as low as 10 s of nVs). Such a precision is not possible with instruments utilizing and measuring DC signals and it is only possible using AC signals thanks to the phase locking technique which has led to the development of low noise lock-in amplifiers.

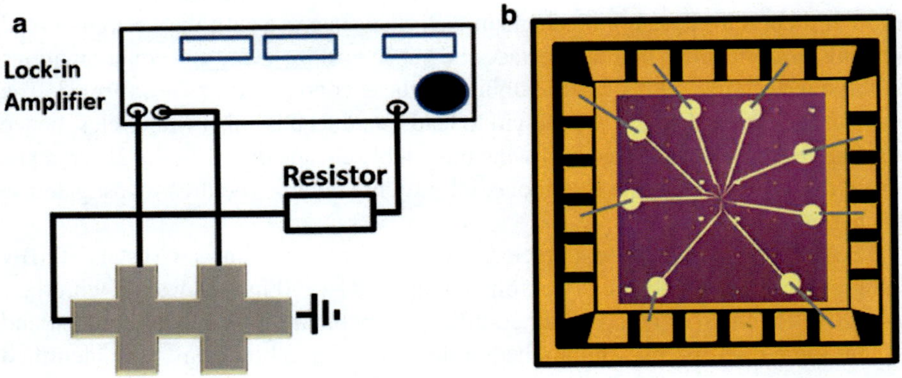

Fig. 3.2 **a** Schematic illustrating four-probe measurement configuration using an AC current (10–100 nA) generated from a lock-in amplifier using a resistor (1–10 MΩ). The frequency of AC excitations used were limited to below 50 Hz to avoid mains hum. **b** An optical micrograph of a typical device consisting of large contact pads enabling wire-bonding onto a 24-pin ceramic leadless chip-package with gold pads

Figure 3.2 shows a schematic of a typical measurement circuit, where the Hall bar device is wire bonded onto a 24-pin ceramic chip package, used to carry out both transverse and longitudinal resistance measurements using AC excitation generated by lock-in amplifier. The function of the resistor (typically between 1 and 10 MΩ) in the circuit is to utilize the AC signal generated by the lock-in amplifier as a constant AC current source. The operating principle of the lock-in amplifier will be outlined in Sect. 3.2.1.

3.2.1 Phase Sensitive Detection and Lock in Amplifier

Let us first consider an ordinary AC signal amplifier and compare it to a lock in amplifier. An attempt to measure a very low voltage (10 nV) and low frequency AC signal (100 Hz) using an ordinary amplifier, which intrinsically comes along with a gain and a bandwidth which for illustrative purposes could be assumed to be 1000 and 1 kHz respectively. Using a state of the art of amplifier we would obtain a noise level of 5 $nV/\sqrt{Hz}$, which would lead to an amplified signal of 10 μV, with a broadband noise of 15.8 mV, which means that our attempted measurement will be overwhelmed by the present noise. Even if we were to use a band pass filter on our amplified signal, to reduce its bandwidth, let's say to 1 Hz, we would still end up with a noise level of 5 μV, which is on the order of our signal that we have amplified. The advantage and in a way the function of the lock in amplifier is to reduce the bandwidth of the measured signal as low as 0.01 Hz, through its phase sensitive detection mechanism, which in the above example would lead to a noise level of 0.5 μV significantly smaller than our signal.

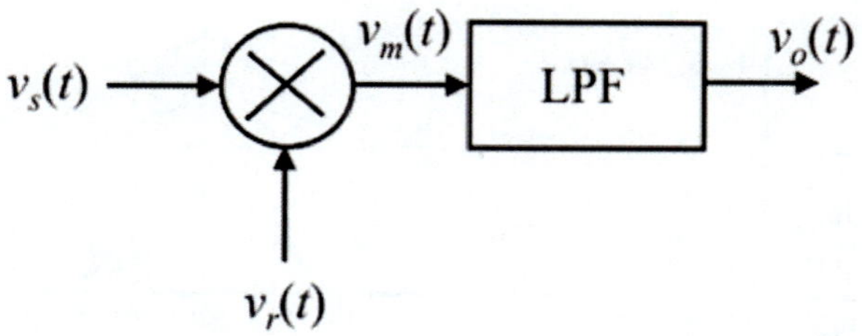

Fig. 3.3 Phase sensitive detector operation principle. Measured signal is multiplied with a reference signal before passing through a low-pass filter filtering out high frequency components. Figure courtesy of T. Tomson

The way the phase sensitive detection mechanism of a lock-in amplifier works could be illustrated through the following simple example. Imagine a reference signal which has a locked frequency to the signal being measured at a fixed phase offset. The signal, which is being measured is in the form of

$$v_s(t) = A_s \cos(\omega_S t + \phi_S) \tag{3.1}$$

where ω_S is the frequency and ϕ_S is the phase. The reference signal supplied by the PSD is of the form

$$v_r(t) = A_r \cos(\omega_r t + \phi_r) \tag{3.2}$$

with its own frequency and phase. The simplistic PSD system considered here carries out the operation in Fig. 3.3, before passing the signal through a low pass filter.

The multiplied signal is mathematically expressed as

$$v_m(t) = \frac{1}{2} A_r A_s [\cos[(\omega_r + \omega_s)t + (\phi_r + \phi_s)] + \cos[(\omega_r - \omega_s)t + (\phi_r - \phi_s)]] \tag{3.3}$$

where the low pass filter used has a cut off frequency $\omega_c \ll \omega_r + \omega_s$ which means that the first term in the brackets in the equation above will be cancelled once the signal goes through the low pass filter. When the term $\omega_r - \omega_s$ is zero, which is the case, and is the main operation principle of phase sensitive detector, one obtains the filtered wave v_0 as

$$v_0 = \frac{1}{2} A_r A_s \cos(\Delta\phi) \tag{3.4}$$

assuming a low unity low pass filter gain. In the event of a constant phase difference between the measured signal and the reference signal one obtains a DC signal proportional to the measured signal amplitude. The PSD could be regarded as transition of the operation of an ordinary low pass filter such that its frequency response is centred on the reference frequency supplied as shown on Fig. 3.4. Hence, when carrying out

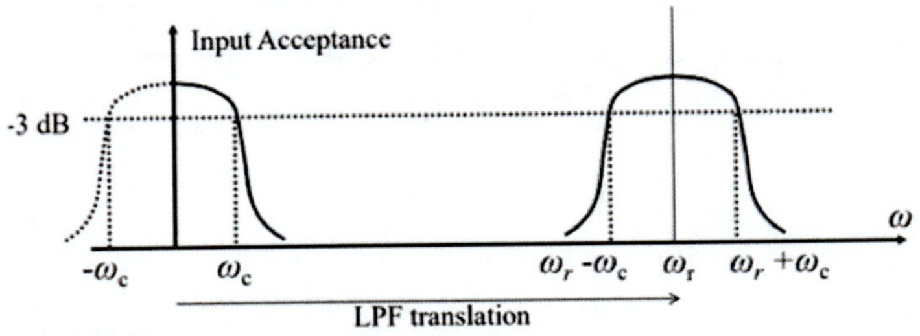

Fig. 3.4 Lock in amplifier enables an effective low pass filter operation with a cut-off frequency centred around the reference signal supplied by the lock in. Figure courtesy of T. Tomson

the electrical measurement schematically depicted in Fig. 3.2a, by sourcing a constant AC current through the resistor, one can measure the amplitude of the voltage drop across the sample using the lock in amplifier, hence obtain the resistance of the sample.

It must be noted that when performing superconductivity measurements, often additional RC filters are used to filter out additional harmonics unavoidably generated by the lock-in amplifiers, which start to have a pronounced influence on the signal once the detected signal quantities become extremely small.

3.3 Measurement Geometry and Electrostatic Gating

Electrical measurements reported in this thesis were carried out using various measurement configurations on Hall bars devices of hBN/RG/hBN van der Waals stacks on Si/SiO_2 substrates as shown in Fig. 3.5. The more usual 'local' measurement geometry depicted in Fig. 3.5b, allows one to determine longitudinal resistance, and the longitudinal resistivity as

$$R_{xx} = \frac{V_{xx}}{I_{SD}} = \frac{\rho_{xx} W}{L} \tag{3.5}$$

with W being the width of the Hall bar and L being the distance between the contact pair used to measure the longitudinal voltage drop V_{xx}. Measuring it in combination with the Hall resistance, as highlighted already in Chap. 2, allows one to extract the carrier density associated with the Fermi level of the system as

$$R_{xy} = \rho_{xy} = -\frac{B}{en} \tag{3.6}$$

where a combination of the two measurements allows one to extract Hall mobility of the system at around given chemical potential.

As well as enabling 4-probe measurements which allows one to obtain resistance of the sample without contact resistance influence (depicted in Fig. 3.5b) Hall bars

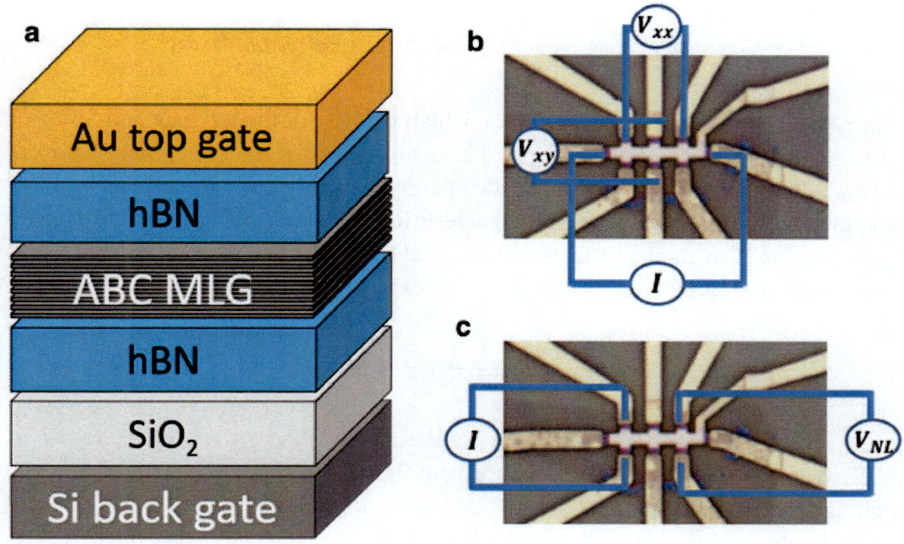

Fig. 3.5 **a** Schematic of the van der Waals heterostructure formed using an all dry transfer process. **b** Local measurement geometry depicted on a double gated device with the current injection, longitudinal and transverse voltage contacts labelled. **c** Non-local measurement geometry with voltage measured away from current injection contacts

fabricated have also enabled the study of topological currents which lead to a non-local voltage drop away from current injection contacts (as depicted in Fig. 3.5c) due to presence of a zero magnetic field Hall effect of spins and or valleys. The presence of non-local signal is verified through an estimation of Ohmic contribution to non-local resistance at the point the non-local voltage measured as [19, 20]

$$R_{Ohmic} = \rho_{xx} e^{-\pi L / W} \tag{3.7}$$

for current applied across the width W voltage measured at length L away from current injection contacts.

An important tuneable parameter while investigating electronic properties of a two-dimensional or a thin film system is the chemical potential. To be able to have a comprehensive insight one needs to be able to tune the Fermi level at each surface of a rhombohedral graphite film. This is done through electrostatic gating by application of a DC voltage across the sample and a gate, through a dielectric located above/below the sample. The top and back gates used within the devices for which the measurements are reported were gold metal contact and n-doped silicon, respectively. The generated electric field across hBN and SiO_2 dielectrics produce a finite carrier density at each surface proportional to voltage applied, which is expressed by the equation below per unit surface area as

$$n_{total} = \frac{V_b \varepsilon_{rb} \varepsilon_0}{ed_b} + n_{b0} + \frac{V_t \varepsilon_{rt} \varepsilon_0}{ed_t} + n_{t0} \quad (3.8)$$

where d is the thickness of the dielectric which is hBN for the top gate and SiO_2 for the bottom gate, and ε_r being the relative permittivity for the dielectrics, with n_{b0} and n_{t0} taking care of the position of the charge neutrality point at each surface. By inducing charges of opposite types on each surface, or applying voltages of opposite signs one can also induce a displacement field to the low energy flat bands of a rhombohedral graphite system which can be expressed as

$$D = \frac{\varepsilon_{rt}(V_t - V_{t0}) - \varepsilon_{rb}(V_b - V_{b0})}{2} \quad (3.9)$$

in terms of respective gate voltages with V_{t0} and V_{b0} being the voltage values of the charge neutrality point. The displacement field in principle should lead to a gap opening at the low energy bands of a rhombohedral graphite system above a critical displacement field [21], as it has been observed in bilayer systems [22–25].

3.4 Temperature Control

3.4.1 Sorption Pump-Controlled Helium-3 Cooling

Majority of the experiments reported in this thesis were carried out utilizing variable temperature inserts (VTI) immersed in liquid helium (^{4}He). Liquid helium has a boiling point of 4.2 K. However, at low enough pressures vaporised ^{4}He exists down to temperatures of around 1 K (see Fig. 3.6a). Temperatures down to 1.4 K were

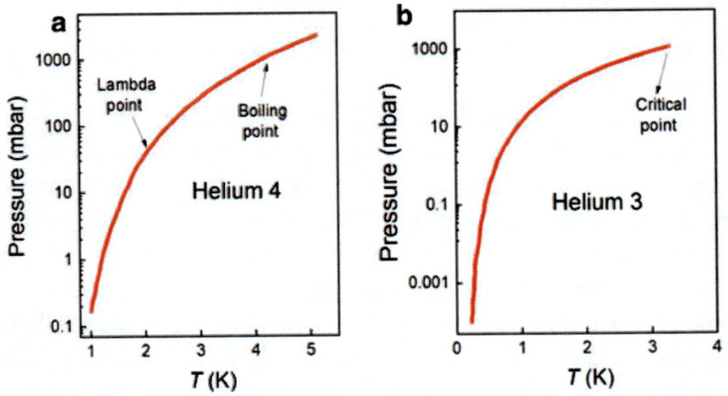

Fig. 3.6 Vapour pressure of **a** ^{4}He and **b** ^{3}He and associated temperatures. At extremely low pressures gaseous ^{3}He exists at temperatures down to 0.25 K whereas only around 1 K is accessible for the case of ^{4}He. Data from Ref. [26]

obtained by pumping vapour helium through the sample space using a rotary oil pump were the vapour pressure was adjusted down to values of around 3 mbar (see Fig. 3.6a) through use of a needle valve.

Temperatures that are not accessible with ^{4}He were obtained using a single shot sorption pumped ^{3}He VTI manufactured by Oxford Instruments. A pictorial image of the VTI is shown in Fig. 3.7a as well as a schematic depicting its operating principle. It can be seen from Fig. 3.7b that extremely low pressures, 0.0001 mbar, are required to reach down to the base temperature of 0.25 K on a ^{3}He system. This is achieved by using a large surface area (which is charcoal in our case) material as sorption pump [26], which the ^{3}He gases adsorb onto.

The operation procedure of the ^{3}He insert depicted in Fig. 3.7a is as follows. The sample which is bonded to a 24-pin chip carrier is initially mounted onto a probe which is attached to end of a narrow finger-like copper block with threads at one end. Using the threads, the copper block is screwed onto the copper cylinder depicted as ^{3}He pot on Fig. 3.7a. Once this procedure is completed, the vacuum can is attached to the insert as an inner vacuum generated through pumping. Once low enough pressure is reached (typically on the order of 10^{-6} mbar), the pumping is terminated and at this stage small amount of exchange (^{4}He) gas ~10 cm^3 is added,

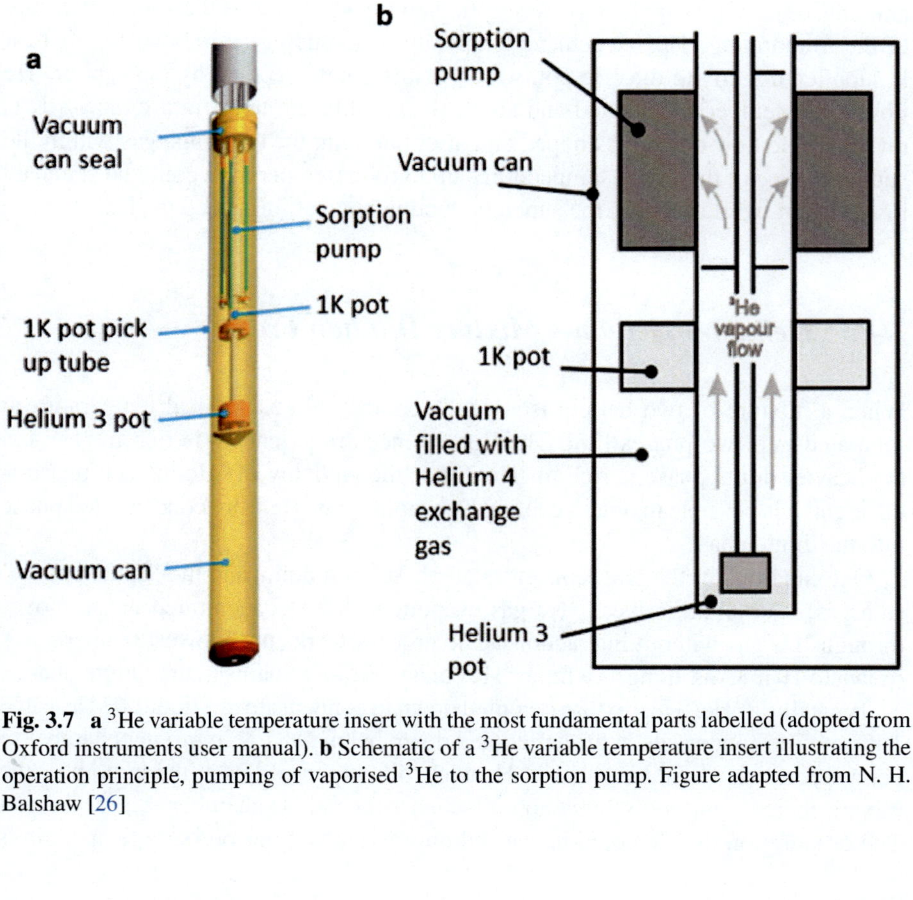

Fig. 3.7 **a** ^{3}He variable temperature insert with the most fundamental parts labelled (adopted from Oxford instruments user manual). **b** Schematic of a ^{3}He variable temperature insert illustrating the operation principle, pumping of vaporised ^{3}He to the sorption pump. Figure adapted from N. H. Balshaw [26]

to ease the cooling process. The VTI is then dipped into a liquid nitrogen Dewar to pre-cool it down to 77 K. Once the insert is cooled to 77 K, it is then carefully dipped into the liquid helium Dewar which if left alone will cool the insert down to 7–8 K (depending on the quality of the vacuum and volume of the exchange gas). At this stage, pumping is started to generate ^{4}He vapour flow through the VTI including the 1 K pot, which will be cooled down to temperatures as low as 1.4 K. A key property of the sorption pump is that, at liquid ^{4}He boiling temperature of 4.2 K it starts absorbing the ^{3}He gas present in the insert, which is what enables it to be used as a pump. Typically, before operation at base temperature of ~0.25 K, using heaters mounted on the sorption pump, it is heated to a temperature above 32 K, while the 1 K pot is kept at the lowest temperature possible (below superfluidity transition point of vaporised ^{4}He at 2.17 K). This releases all ^{3}He atoms adsorbed to the charcoal surface, generating a high pressure, and as the ^{3}He atoms are passing by the 1 K pot they become condensed and fill the ^{3}He pot in a liquid state. This process is referred to as regeneration. Once regeneration process is completed, heater of the sorb is switched off, which cools it down to 4.2 K, and pumping begins with low enough vapour pressures such that surroundings are cooled to temperatures as low as 0.25 K.

Typically, to achieve temperatures between 0.25 and 1.5 K within the VTI, varying amounts of heat are supplied to the sorption pump which varies the vapour pressure of the evaporating ^{3}He. To achieve temperatures including and above 1.6 K, heat is supplied directly to the ^{3}He pot, which at this point is cooled by the vapour ^{4}He flow. Once temperatures equal and above the equilibrium temperature of 7–8 K is reached, ^{4}He flow could be stopped and after pumping the exchange gas within the vacuum space of the insert, temperatures up to room temperature could be achieved inside insert, while keeping the superconducting magnet immersed in ^{4}He.

3.4.2 Helium-3/Helium-4 Mixture Dilution Fridge

When a mixture of two helium isotopes is cooled below a critical temperature it separated into two phases [26]. The light 'concentrated phase' is rich in ^{3}He and the heavier dilute phase is rich in ^{4}He. Since the enthalpy of ^{3}He in each phase is different, it is possible to obtain cooling by evaporating ^{3}He from concentrated phase into the dilute phase.

One can think of the ^{3}He 'concentrated phase' as a liquid and the 'diluted phase' as a gas. As helium is inert, ^{3}He gas evaporated from concentrated phase moves through ^{4}He gas without interactions. The process works until lowest temperatures (down to 10 mK) as along as a finite ^{3}He concentration remains in the 'dilute phase'.

Typically, a ^{3}He/^{4}He mixture is cooled down to condensate in a liquid ^{4}He assisted 1 K pot. The condensate is further cooled down below 0.87 K in mixing chamber to get a separation of the two phases. Heavier 'dilute phase' lies at the bottom of the mixing chamber and cools the sample attached to the mixing chamber as it evaporates. The evaporation of ^{3}He could be carried out to yield vapour pressure temperatures beyond that of the charcoal pump of the ^{3}He sorption pump system hence leading to an order of magnitude lower temperatures. The evaporated ^{3}He/^{4}He mixture is stored in the dilution fridge to be sent back to 1 K pot for the next cooling cycle.

3.5 Magnetic Field

Magneto-transport experiments reported in this thesis takes advantage of the extremely high magnetic fields (up to 18 T) generated in superconducting magnets immersed in liquid helium [26]. The main motivation behind superconducting magnet technology is that their property of zero resistance allows them to behave as an ideal inductor without resistance.

A superconducting magnet typically consists of a solenoid made from superconducting metals such as NbTi and NbSn and its magnetic field generation and basic properties can be understood using introductory level inductance physics. As an example, we can consider a simple solenoid with zero resistance carrying current i with N number of coils and a cross section area of A, where the generated magnetic flux would be

$$\phi_B = BA = \mu_0 N i A \tag{3.10}$$

with μ_0 being the vacuum permeability where the inductance of the wire could be defined as

$$L = \frac{\phi_B}{i} \tag{3.11}$$

which defines the flux generation efficiency of the solenoid per unit current applied. The most important property of an inductor is that there is only a finite potential drop across it (let us say from point a to b across the inductor), when there is a change of the current applied through it

$$V_{ab} = L\frac{di}{dt} \tag{3.12}$$

as expressed in the equation above. We note that there is a self-induced electromotive force within the inductor that opposes this change of applied current. Hence, a finite energy input to the solenoid is necessary to be able to generate finite final current I flowing through it. The instantaneous power applied to the inductor is expressed as

$$P = V_{ab} i = Li\frac{di}{dt} \tag{3.13}$$

hence one can write the total energy applied to the inductor to be able to generate a final current of I as

$$U = L\int_{0}^{I} i di = \frac{1}{2} L I^2 \quad (3.14)$$

so an inductor with a finite current I flowing through it has a finite energy stored which is released as this current decreased.

The energy stored within a superconducting magnet may be rapidly released in a quenching event when a fraction of the superconductor switches to a normal state (which could be due to a variety of reasons such as quick sweeping of current or too high magnetic field) which kicks off a chain reaction at which the entire superconducting magnet switches to a metallic state releasing the entire energy stored within it as heat to the surroundings, which may cause significant damage to the equipment.

A superconducting magnet system is typically equipped with a superconducting switch (shown in Fig. 3.8), which is a narrow strip of superconducting (below a Tc accessible by switch heater) metal, connected to a switch heater which keeps this narrow strip of wire in a resistive state as the magnet is charged with energy. The function of the superconducting switch is to enable the operation of the magnet in a persistent mode, which makes use of the constant field generated inside the magnet once a given current is applied. Typically, once the given value of magnetic field is reached, the switch heater is turned off, forming a superconducting circuit, and passivating the power supply. This approach used to reduce the heat load on the cryostat enabling the magnet power supply to be turned off. Another advantage is the stability of the field generated for experiments that strictly require a constant

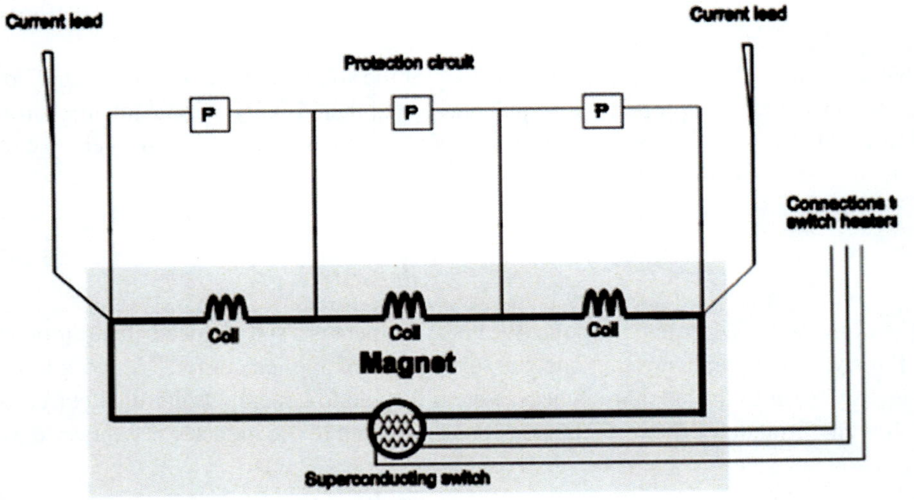

Fig. 3.8 A superconducting magnet circuit (grey shaded area) showing the superconducting coils as well as superconducting switch connected to a heater. The protection circuit (labelled) is necessary for minimising the energy release during potential quenching events. Figure adapted from N. H. Balshaw [26]

magnetic field. The persistent magnetic field generated at a given current across the magnet is typically more stable than the power supply as small fluctuations in current occurs within, leading a small uncertainty in the magnetic field value.

3.6 Summary

Rhombohedral graphite flakes identified through Raman spectroscopy mapping were encapsulated with high quality hBN crystals using a PMMA/PDMS assisted all-dry transfer method. Hall bars were then defined on the formed van der Waals heterostructures with gold contacts as well as a top gate by a combination of e-beam lithography, etching and e-beam evaporation processes.

Low noise electrical measurements were carried out using SR830 lock-in amplifiers whose AC signals were sometimes superimposed onto a DC signal enabling direct dV/dI measurements. Kiethley sourcemeters were used DC signal sources. Prior to measurements the devices fabricated on SiO_2/Si wafers were cleaved and wire bonded onto 24-pin ceramic chip carriers that were carefully mounted onto probes attached to the liquid helium variable temperature inserts.

Lastly, magnetic field generation (up to 18 T) and control was enabled through use of superconducting magnet technology. During the experiments instrument control data acquisition was carried out using LabVIEW programming language.

References

1. Novoselov KS et al (2004) Electric field effect in atomically thin carbon films. Science 306(5696):666–669. https://doi.org/10.1038/nmat1849
2. Novoselov KS et al (2005) Two-dimensional gas of massless Dirac fermions in graphene. Nature 438(7065):197–200. https://doi.org/10.1038/nature04233
3. Zhang Y, Tan YW, Stormer HL, Kim P (2005) Experimental observation of the quantum Hall effect and Berry's phase in graphene. Nature 438(7065):201–204. https://doi.org/10.1038/nature04235
4. Hwang EH, Adam S, Das Sarma S (2007) Carrier transport in two-dimensional graphene layers. Phys Rev Lett 98(18):2–5. https://doi.org/10.1103/PhysRevLett.98.186806
5. Ishigami M, Chen JH, Cullen WG, Fuhrer MS, Williams ED (2007) Atomic structure of graphene on SiO_2. Nano Lett 7(6):1643–1648. https://doi.org/10.1021/nl070613a
6. Morozov SV et al (2008) Giant intrinsic carrier mobilities in graphene and its bilayer. Phys Rev Lett 100(1):11–14. https://doi.org/10.1103/PhysRevLett.100.016602
7. Chen JH, Jang C, Xiao S, Ishigami M, Fuhrer MS (2008) Intrinsic and extrinsic performance limits of graphene devices on SiO_2. Nat Nanotechnol 3(4):206–209. https://doi.org/10.1038/nnano.2008.58
8. Du X, Skachko I, Barker A, Andrei EY (2008) Approaching ballistic transport in suspended graphene. Nat Nanotechnol 3(8):491–495. https://doi.org/10.1038/nnano.2008.199
9. Novoselov KS et al (2005) Two-dimensional atomic crystals. Proc Natl Acad Sci U S A 102(30):10451–10453. https://doi.org/10.1073/pnas.0502848102

10. Watanabe K, Taniguchi T, Kanda H (2004) Direct-bandgap properties and evidence for ultraviolet lasing of hexagonal boron nitride single crystal. Nat Mater 3(6):404–409. https://doi.org/10.1038/nmat1134
11. Giovannetti G, Khomyakov PA, Brocks G, Kelly PJ, Van Den Brink J (2007) Substrate-induced band gap in graphene on hexagonal boron nitride: ab initio density functional calculations. Phys Rev B Condens Matter Mater Phys 76(7):2–5. https://doi.org/10.1103/PhysRevB.76.073103
12. Dean CR et al (2010) Boron nitride substrates for high-quality graphene electronics. Nat Nanotechnol 5(10):722–726. https://doi.org/10.1038/nnano.2010.172
13. Dean CR et al (2011) Multicomponent fractional quantum Hall effect in graphene. Nat Phys 7(9):693–696. https://doi.org/10.1038/nphys2007
14. Mayorov AS et al (2011) Micrometer-scale ballistic transport in encapsulated graphene at room temperature. Nano Lett 11(6):2396–2399. https://doi.org/10.1021/nl200758b
15. Castellanos-Gomez A et al (2014) Deterministic transfer of two-dimensional materials by all-dry viscoelastic stamping. 2D Mater 1(1). https://doi.org/10.1088/2053-1583/1/1/011002
16. Uwanno T, Hattori Y, Taniguchi T, Watanabe K, Nagashio K (2015) Fully dry PMMA transfer of graphene on h-BN using a heating/cooling system. 2D Mater 2(4). https://doi.org/10.1088/2053-1583/2/4/041002
17. Novoselov KS, Mishchenko A, Carvalho A, Castro Neto AH (2016) 2D materials and van der Waals heterostructures. Science 313. https://doi.org/10.1126/science.aac9439
18. Latychevskaia T et al (2019) Stacking transition in rhombohedral graphite. Front Phys 14(13608). https://doi.org/10.1007/s11467-018-0867-y
19. Balakrishnan J, Kok Wai Koon G, Jaiswal M, Castro Neto AH, Özyilmaz B (2013) Colossal enhancement of spin-orbit coupling in weakly hydrogenated graphene. Nat Phys 9(5):284–287. https://doi.org/10.1038/nphys2576
20. Gorbachev RV et al (2014) Detecting topological currents in graphene superlattices. Science 346(6208):448–451. https://doi.org/10.1126/science.1254966
21. Koshino M (2010) Interlayer screening effect in graphene multilayers with ABA and ABC stacking. Phys Rev B Condens Matter Mater Phys 81(12):1–7. https://doi.org/10.1103/PhysRevB.81.125304
22. Castro EV et al (2007) Biased bilayer graphene: semiconductor with a gap tunable by the electric field effect. Phys Rev Lett 99(21):8–11. https://doi.org/10.1103/PhysRevLett.99.216802
23. Oostinga JB, Heersche HB, Liu X, Morpurgo AF, Vandersypen LMK (2008) Gate-induced insulating state in bilayer graphene devices. Nat Mater 7(2):151–157. https://doi.org/10.1038/nmat2082
24. Zhang Y et al (2009) Direct observation of a widely tunable bandgap in bilayer graphene. Nature 459(7248):820–823. https://doi.org/10.1038/nature08105
25. Taychatanapat T, Jarillo-Herrero P (2010) Electronic transport in dual-gated bilayer graphene at large displacement fields. Phys Rev Lett 105(16):1–4. https://doi.org/10.1103/PhysRevLett.105.166601
26. Balshaw N (1996) Practical cryogenics. An introduction to laboratory cryogenics. Oxford Instruments Superconductivity Limited

Chapter 4
Bulk Versus Surface Conduction in Rhombohedral Graphite Films

4.1 Introduction

Bulk nodal line semimetal [1] rhombohedral graphite is a topological insulator in thin film limit as highlighted already in Chap. 1. Albeit bearing similarities to more usual topological Z2 symmetry protected topological insulators [2], rhombohedral graphite thin films are chiral symmetry protected and are a 3D generalisation of the simplest 1D topological insulator model Su-Schrieffer-Heeger chain [3–5]. Nevertheless, both systems possess topological surface states which lie inside their bulk band gaps [6].

A fundamental issue to transport studies of topological insulator crystals has been the unavoidable albeit limitable parallel thermally activated bulk conduction [7–9]. These crystals require careful crystal growth under laboratory conditions to unleash their topological properties, unlike rhombohedral graphite films which are exfoliated from high quality crystals of naturally occurring graphite [10–12].

The bulk band gap of rhombohedral graphite films, in the thicknesses of the films studied (9 to 50 layers) with tight binding predicted bulk gap sizes ranging from 370 to 71 meV is comparable to bulk gap sizes of ordinary topological insulators. During experiments, one would expect a smaller transport gap size to manifest itself as it has been the case in other topological insulators [7], where one would hope, the conduction at low temperature at least, to be dominated by the surface states of topological origins.

4.2 Temperature Dependence of Zero Gate Resistivity

The temperature dependence of zero gate resistivity of rhombohedral graphite films with thicknesses ranging from 10 to 21 layers is shown in Fig. 4.1 as well as a reference 10-layer Bernal hexagonal stacked graphite device. Until a saturation temperature between 50 and 75 K, the resistivity of rhombohedral graphite devices increases

S. Ozdemir, *Electronic Properties of Rhombohedral Graphite*, Springer Theses,
https://doi.org/10.1007/978-3-030-88307-2_4

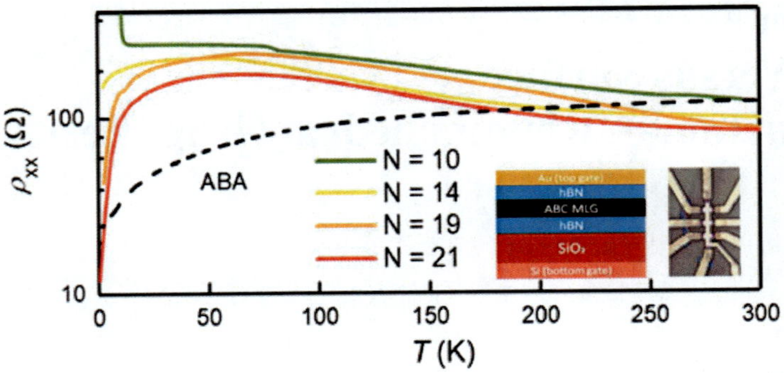

Fig. 4.1 Temperature dependence of resistivities of Hall bars devices rhombohedral graphite films of 10, 14, 19 and 21 layers thickness as well as a 10-layer thick reference hexagonal graphite film device with a heterostructure schematic as well as a device micrograph insets

with decreasing temperature, as it would be expected from an intrinsic semiconductor, whereas for the case of reference ABA device, resistivity is decreasing, as it would be expected from a semimetal. We know that, in low energy limit, rhombohedral graphite is also a semimetal thus the observed semiconductor like cooling curves could only be explained an expected thermally activated bulk conduction domination at temperatures above 50 K. The drop in resistance at low temperatures (except the 10 layer device—for which a gap at charge neutrality point emerges which is discussed in detail on Chap. 6) is parallel surface conduction channels emerging, and in fact, dominating the transport properties.

Extracted carrier densities (from Hall resistance) for a 14-layer rhombohedral graphite film is shown on Fig. 4.2a. As expected from thermally activated conduction on an intrinsic semiconductor, bulk carrier density increases exponentially with temperature. The Arrhenius plot of the temperature dependence of carrier densities (Fig. 4.2b) was found to yield a thermal activation gap of (110 ± 20) meV which

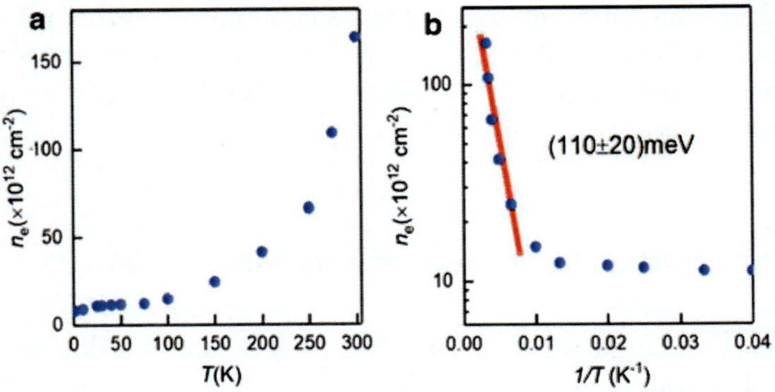

Fig. 4.2 **a** Temperature dependence of Hall resistance extracted electron carrier densities of a 14-layer rhombohedral graphite film **b** Arrhenius plot of the carrier densities plotted in **a** with an obtained activation gap size of (110 ± 20) meV

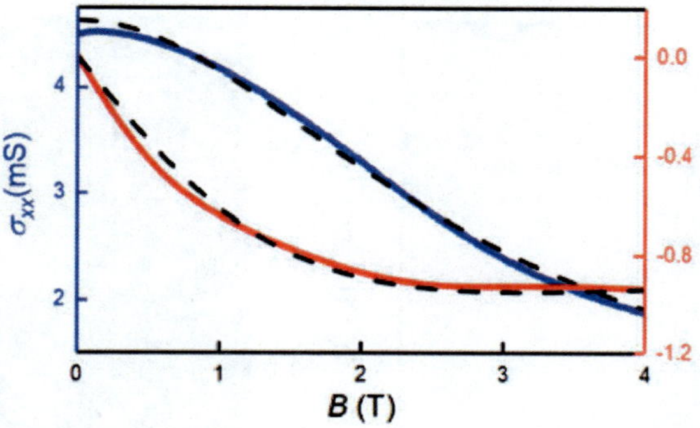

Fig. 4.3 Multiband transport on 14-layer rhombohedral graphite film at 5 K with longitudinal and transverse conductivities (blue and red) both described by the same 3 carrier model (dashed lines), yielding 2 surface electrons and holes, and one bulk (electrons) carriers

is as expectedly smaller than theoretically calculated bulk gap size of 245 meV [5]. The extracted activation gap is comparable to the ones for high quality $BiSbTe_2S$ topological insulator crystals [13].

4.3 Multi Carrier-Type Transport

To clarify for the claimed multiband conduction, taking place across surface and bulk bands at low temperature, we have carried out multiple carrier fits (Fig. 4.3) by carrying out a list-squares fitting of the expressions in Eq. (2.39) to longitudinal and transverse conductivities, σ_{xx} and σ_{xy} which was found to suggest 3 types of carrier at low temperatures (5 K) where we expect surface conduction to dominate over that of bulk. We attribute two carriers of opposite signs with relatively higher mobilities of $3420/3900\frac{\text{cm}^2}{\text{Vs}}$ and carrier densities of $(2.11/4.12) \times 10^{12}\ \text{cm}^{-2}$ to surface charges of electrons and holes respectively which emerge and increasingly become dominant as temperature is decreased. The remaining charge type are electrons which we believe to be of bulk origin with a mobility of $1040\frac{\text{cm}^2}{\text{Vs}}$ and a carrier density of $5.38 \times 10^{12}\ \text{cm}^{-2}$ which completely dominates transport at higher temperatures due to increased thermal activation across the bulk gap.

4.4 Semiconductor—Metal Transition

Remaining evidence for crossover from a semiconductor like bulk transport to metallic surface transport are the on the top and back gate maps at around the crossover point of $T = 50$ K and a surface dominated conduction temperature of $T = 0.25$ which are shown in Fig. 4.4a, b. The contrast between the two maps high-

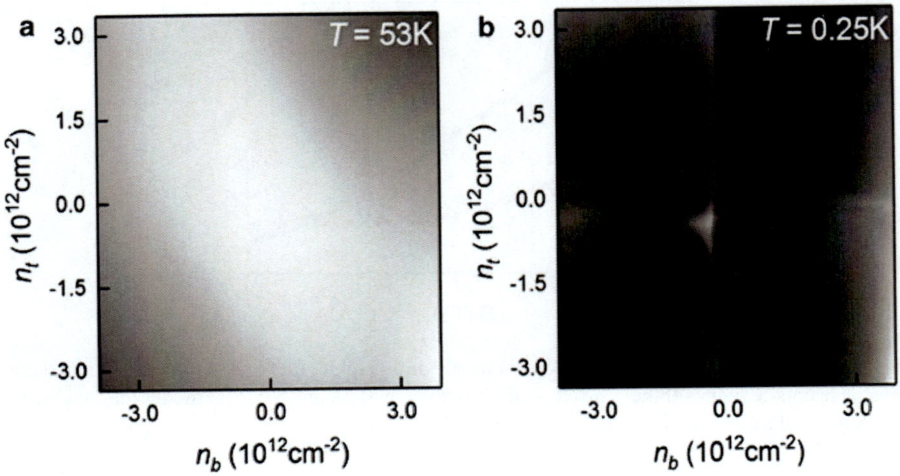

Fig. 4.4 Resistivity maps of the 14-layer rhombohedral graphite device with top and bottom surface carrier densities tuned through capacitive gating **a** at T = 53 K and **b** at T = 0.25 K. The colour scale on both maps corresponds to 240–40 Ω, white to black

lights the transition each sample undergoes as cooled to temperatures down to and below T = 0.25 K. It can be seen on Fig. 4.4a that the maximum resistivity points as expected are located diagonally along the displacement field line at T = 53 K, whereas a cross feature is visible at T = 0.25 K indicating absence of a single semiconductor thin film but rather presence of a thin film with two electronically decoupled metal surfaces. We note that, albeit conduction at this point being dominated by surface states, there are still a significant number of bulk charges, which we will discuss the transport properties of in Chap. 8.

4.5 Summary

Conduction properties of rhombohedral graphite films, as well as topological surface states, possesses gapped bulk bands, across which a thermally activated bulk conduction takes place and dominates transport properties of rhombohedral graphite films above 50 K. The size of this transport gap is found to be (110 ± 20) meV for a 14-layer film, considerably less than theoretically expected value of 245 meV. This thermally activated bulk conduction is exponentially supressed with decreasing temperature, hence leading to a surface state dominated transport accompanied by a semiconductor-metal transition at low temperatures. As a result, at 5 K and below, rhombohedral graphite films exhibit multi-carrier type transport, with surface electron and hole charges as well as bulk charges. The transition from bulk to surface dominated transport occurring with decreasing temperature is also evidenced with top and back gate maps where carrier densities on each surface are tuned independently around the charge neutrality point.

References

1. Heikkilä TT, Volovik GE (2011) Dimensional crossover in topological matter: evolution of the multiple Dirac point in the layered system to the flat band on the surface. JETP Lett 93(2):59–65. https://doi.org/10.1134/S002136401102007X
2. Fu L, Kane CL, Mele EJ (2007) Topological insulators in three dimensions. Phys Rev Lett 98(10):1–4. https://doi.org/10.1103/PhysRevLett.98.106803
3. Xiao R, Tasnádi F, Koepernik K, Venderbos JWF, Richter M, Taut M (2011) Density functional investigation of rhombohedral stacks of graphene: Topological surface states, nonlinear dielectric response, and bulk limit. Phys. Rev. B-Condens. Matter Mater. Phys. 84(16):1–15. https://doi.org/10.1103/PhysRevB.84.165404
4. Asbóth JK, Oroszlány L, Pályi A (2016) The Su-Schrieffer-Heeger (SSH) model. In: A short course on topological insulators: band structure and edge states in one and two dimensions. Springer International Publishing, Cham, pp 1–22
5. Slizovskiy S, McCann E, Koshino M, Fal'ko VI (2019) Films of rhombohedral graphite as two-dimensional topological semimetals. Commun. Phys. 2(1):1–10. https://doi.org/10.1038/s42005-019-0268-8
6. Ho CH, Chang CP, Lin MF (2016) Evolution and dimensional crossover from the bulk subbands in ABC-stacked graphene to a three-dimensional Dirac cone structure in rhombohedral graphite. Phys Rev B 93(7). https://doi.org/10.1103/PhysRevB.93.075437
7. Ren Z, Taskin AA, Sasaki S, Segawa K, Ando Y (2010) Large bulk resistivity and surface quantum oscillations in the topological insulator Bi2 Te2 Se. Phys Rev B-Condens Matter Mater Phys 82(24):1–4. https://doi.org/10.1103/PhysRevB.82.241306
8. Skinner B, Chen T, Shklovskii BI (2012) Why is the bulk resistivity of topological insulators so small? Phys Rev Lett 109(17):1–5. https://doi.org/10.1103/PhysRevLett.109.176801
9. Barreto L et al (2014) Surface-dominated transport on a bulk topological insulator. Nano Lett 14(7):3755–3760. https://doi.org/10.1021/nl501489m
10. Henni Y et al (2016) Rhombohedral multilayer graphene: a magneto-Raman scattering study. Nano Lett 16(6):3710–3716. https://doi.org/10.1021/acs.nanolett.6b01041
11. Yang Y et al (2019) Stacking order in graphite films controlled by van der Waals technology. Nano Lett 19(12):8526–8532. https://doi.org/10.1021/acs.nanolett.9b03014
12. Shi Y et al (2020) Electronic phase separation in multilayer rhombohedral graphite. Nature 584(7820):210–214. https://doi.org/10.1038/s41586-020-2568-2
13. Kushwaha SK et al (2016) Sn-doped Bi1.1Sb0.9Te2S bulk crystal topological insulator with excellent properties. Nat Commun 7(1):11456. https://doi.org/10.1038/ncomms11456

Chapter 5
Landau Level Spectroscopy of Rhombohedral Graphite Films

5.1 Introduction

We have seen that at low temperatures, the transport properties of rhombohedral graphite films are dominated by two-dimensional low energy surface states. One of the most powerful techniques to characterise electronic structure of high-quality two-dimensional systems is to study their magnetic field dependence, including the study of quantum Hall effect and carrying out of Landau level spectroscopy experiments. Hence, this chapter will be dedicated to Landau level spectroscopy of the low energy surface states, with the overall goal of gaining a complete understanding of dispersion properties of low energy surface states of rhombohedral graphite films.

5.2 Single Gated Landau Fan Maps

The initial magnetic field characterisation of the rhombohedral graphite samples, ranging from thicknesses of 9 to 50 layers, was carried out on a single surface which are shown in Fig. 5.1. It can immediately be recognised that there is a significant difference between the experimental data and the magnetic field spectra calculated for rhombohedral graphite films of up to 9 layers in Chap. 1, Sect. 1.12. The most striking feature that is common to all four maps is the strong electron-hole asymmetry. The parameter for electron-hole asymmetry within the tight binding framework is γ_4, [1] which already indicates that simplest Landau level spectrum [2] taking into account the parameters γ_0 and γ_1 solely is not enough and hence a more detailed model is required.

There are also other qualitative features of the maps that have a trend as the layer number of graphene stacks is increased. The electron side of each map (positive doping) there are linear in B-field Landau level features which, at a first glance suggests a relatively simple dispersion on the electron side assuming of course the 4 N zeroth Landau levels are still degenerate as it has suggested to be in previous

S. Ozdemir, *Electronic Properties of Rhombohedral Graphite*, Springer Theses,
https://doi.org/10.1007/978-3-030-88307-2_5

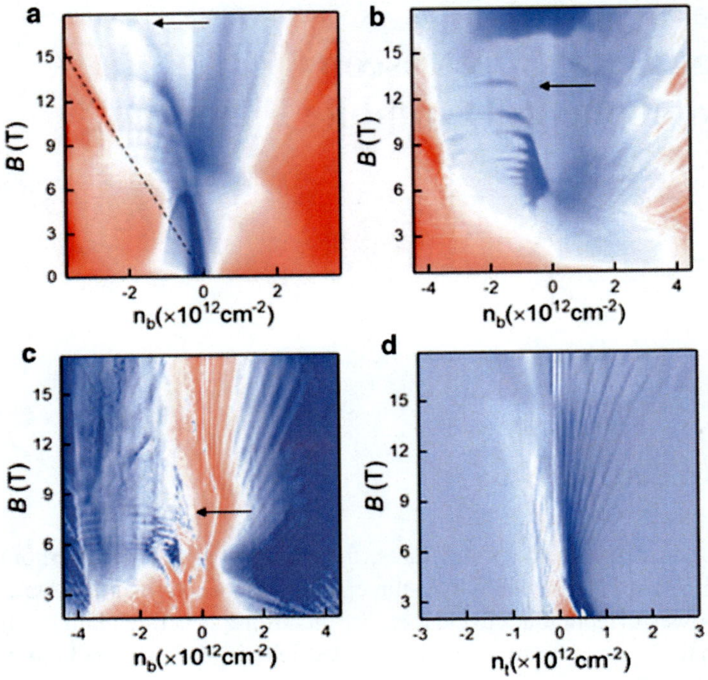

Fig. 5.1 Single gated magnetic field stepped map of **a** longitudinal conductivity of a 9 layer rhombohedral graphite film as a function of bottom surface carrier density (blue to red logarithmic scale bar as 0.2–43 ms), **b** longitudinal conductivity of a 19 layer film (blue to red logarithmic scale as 0.1–500 ms), **c** absolute differential longitudinal conductivity of a 21 layer film as a function of bottom surface carrier density (blue to red logarithmic scale bar as 50 uS/V to 0.5 mS/V), **d** absolute differential longitudinal conductivity of a 50 layer thick film (blue to red scale bar as −0.5–0.5 mS/V) as a function of top surface carrier density

studies. The hole side, however, is very complicated, with horizontal features as well as vertical Landau level features that arise away from zero doping. An interference of these features is clearly visible on the hole side of Fig. 5.1a. The maximum field at which horizontal features persist is 17 T on single gated 3 nm device marked with a black arrow and it is found to drop to 13 and 8 T on 6.5 and 7.2 nm devices respectively with an eventual complete absence of any features on the hole side for the 50 layer device on Fig. 5.1d.

5.3 Implied Low Energy Band Structure

Most of the qualitative features of the Landau fan map on Fig. 5.1a (except the 5–0 T magnetic field feature around charge neutrality point will be considered as a spontaneous quantum phase transition on Chap. 6) was captured by a calculated magnetic field spectrum [3, 4] taking into account of additional hopping parameters

γ_2, γ_3 and γ_4 which can be summarised with the following two band Hamiltonian as

$$H = \begin{pmatrix} \frac{\pi\pi^*}{2m_*} & X(p) \\ X^*(p) & \frac{\pi\pi^*}{2m_*} \end{pmatrix} \tag{5.1}$$

with π and π^* being $p_x - ip_y$ and its complex conjugate respectively with effective mass term m_*^{-1} term being equal to $4v^2\gamma_4/\gamma_1\gamma_0 \cong 0.4m_e$. The diagonal terms are given by

$$X(p) = \sum_{\{n_1,n_2,n_3\}} \frac{(n_1+n_2+n_3)!}{n_1!n_2!n_3!} \frac{1}{(-\gamma_1)^{n_1+n_2+n_3-1}} (v\pi)^{n_1} \left(v_3\pi^*\right)^{n_2} \left(\frac{\gamma_2}{2}\right)^{n_3}$$

and its complex conjugate, respectively. The above two band Hamiltonian yields the low energy band structure of a N-layer rhombohedral graphite film as

$$E(p) = \frac{p^2}{2m_*} \pm |X(p)| \tag{5.2}$$

which is plotted for a 9-layer film in Fig. 5.2 and is the suggested low energy band structure of the hBN encapsulated rhombohedral graphite thin films.

The most notable feature of the band structure apart from trigonal warping accompanied by the quadratic dispersion is the 3 mini-valleys (depicted by black arrows on Fig. 5.2) that form on the valance band as a band overlap of $4\gamma_1\gamma_4/\gamma_0$ develops meaning that any gap generation in these weakly dispersing low energy bands needs to come over this band overlap value.

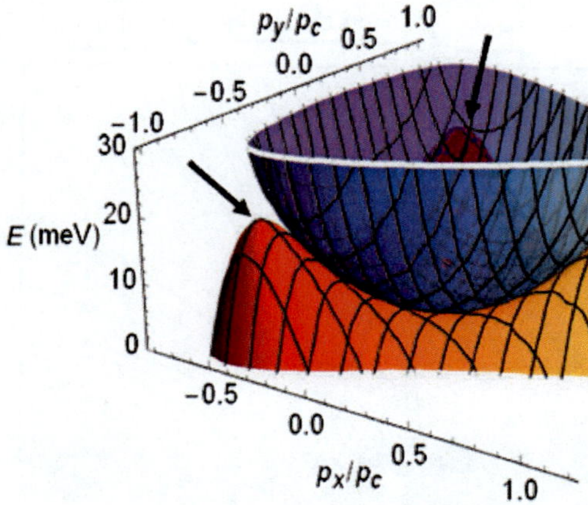

Fig. 5.2 Low energy band structure of a 9-layer rhombohedral graphite film implied by a full tight binding model

5.4 Discussion

The Landau fan spectrum implied by the gapless electronic band structure in Fig. 5.2 is plotted on Fig. 5.3. The most striking feature in contrast to earlier studies is the fact that the degeneracy of N zeroth Landau levels is lifted (see the zeroth Landau levels label on Fig. 5.3), albeit valley and spin degeneracy retained hence leading to 9 individual Landau levels (as valley and spin degeneracy hold) originating from a single surface with particular valleys on each surface. These Landau levels, as suggested from the Landau fan spectrum may account for most of the linear-like dispersion features on the electron side of single gated Landau fan spectra on Fig. 5.1. The hole Landau levels arising as triply degenerate at low magnetic fields, due to three mini valleys formed at the valance band (see Fig. 5.2, black arrows), split at higher magnetic fields, thus forming crossing with zeroth Landau levels explaining the interference effects of features on the hole side on Fig. 5.1d. As well as the interference effects the hole Landau levels are also responsible for the horizontal features which become absent for the 50-layer thin film. The robustness of zeroth Landau levels are likely to be behind the remaining electron side features on 50-layer thin film despite a complete disappearance of any features on the hole side.

One of the features on the hole side of Landau fan maps for 7–11 thick films was a filling factor which was found to coincide with layer numbers estimated from measured flake thickness through atomic force microscopy. This feature, with a ν of -9 for the case of 9-layer device, black dashed line, is visible on Fig. 5.1a. Figure 5.4a–c show respective maps for 7-layer, 9-layer and 11-layer film. The Landau levels can clearly be seen at negative filling factors corresponding to the layer number. We believe that emergence of these filling factors, in the event of single gating which would provide filling factors for a single valley only, hence leading to 2 N zeroth

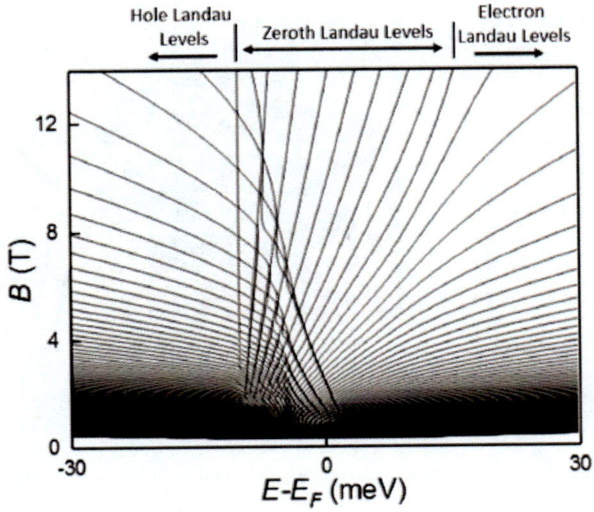

Fig. 5.3 Free particle Landau level spectrum of a 9 Layer rhombohedral graphite film

Landau levels up to ν of ± 9. This is an additional supportive fact to the presence of lifted orbital degeneracy of zeroth Landau levels, hence the full tight binding model leading to the spectrum plotted on Fig. 5.3.

Figure 5.5 shows a double gated Landau fan map of a 9-layer rhombohedral graphite film along the maximum carrier density line of the top—back gate map (both gates tuned simultaneously to induce same density of charges on each surface) hence bringing in the additional valley from the second surface zeroth Landau levels. As expected, robust zeroth Landau levels features up to filling factors of ± 18 are present, with a total of up to 36 levels arising from 9 orbital, 2 spin and 2 valley degrees of freedom aside from a reproduction of the numerous hole Landau level crossings on the hole side. The features of the Landau level spectrum shown in Fig. 5.4a are distinctly clearer than the ones of Fig. 5.1 as it was measured at 10 mK unlike the previous maps, hence leading to less thermal smearing of the levels as well as minimising the thermally activated bulk charges. Features are particularly well pronounced on the whole side at filling factors of $\nu = 8{,}12{,}16$ which can be attributed to larger Zeeman gap between orbital levels. Additional features, apart from the gap at zero doping and the robust $\nu =$ -6 state most likely associated with the gap, which will be considered in detail on Chap. 6, there are the numerous Landau level crossings that take place at an electron doping of $\cong 2.5 \times 10^{12}$ cm^{-2} as illustrated more clearly on Fig. 5.5b. The most striking features of the crossings is that they all occur at even filling factors (labelled by red triangles on Fig. 5.5b) of 10, 14, 18 and 22. The crossings which are only present at 10 mK, suggest emergence of the an additional gap to one that is already present at zero magnetic field, away from zero doping for which valley and spin degrees of freedom may play an equally important role as well as the orbital degree of freedom from each layer. The even filling factors suggests the most likely presence of the quantum spin Hall phase in this regime as depicted on Fig. 5.5d as it enables valley degeneracy as opposite spins are separated. The presence of a layer antiferromagnetic phase would lift valley degeneracy as well as spin, hence lead to crossings at both even and odd filling factors which is why it would be ruled out.

5.5 Summary

Low energy band structure of rhombohedral graphite films was studied through Landau level spectroscopy. The most striking feature of the common to devices of all thicknesses was the strong electron-hole asymmetry. The clearest Landau level maps were obtained on thinner devices (9–10 layers) most probably due to smaller influence of parallel bulk charges due to larger bulk band gaps and less thermal activation. A free particle spectrum calculated for a 9-layer ABC stacked system was found to produce main features of the experimentally obtained Landau fan maps, with the implication being an influence of hopping parameters γ_2, γ_3 and γ_4 within tight binding framework. This meant that the low energy band structure is not dispersing as $E \sim p^N$ but rather as $E \sim p^2$ with a considerable band overlap of the valance

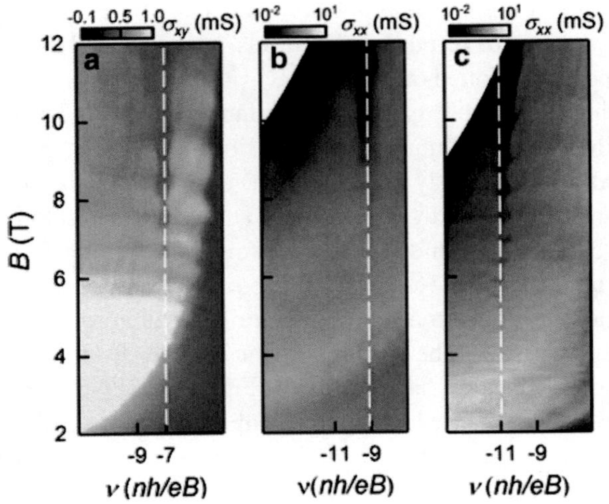

Fig. 5.4 Single gated Landau fan map of longitudinal conductivity σ_{xx} of **a** 7 layer, **b** 9 layer **c** 11 layer device as a function of filling factor ν and magnetic field B, with robust $\nu = -N$ highlighted with dotted white lines

band to that of conduction, where a triply degenerate mini-valley maxima arise due to the trigonal warping of the band structure. The main features of the Landau level spectra that are to be discussed are a lifted degeneracy of the 2 N zeroth Landau levels and numerous Landau level crossings on the hole side arising basically due to the valance-conduction band overlap. Aside from a spontaneous gapped quantum state which will be considered in the next chapter, an additional phase transition takes place on the electron side away from zero doping, most likely to be a transition into a quantum spin Hall phase due to observed filling factor crossings.

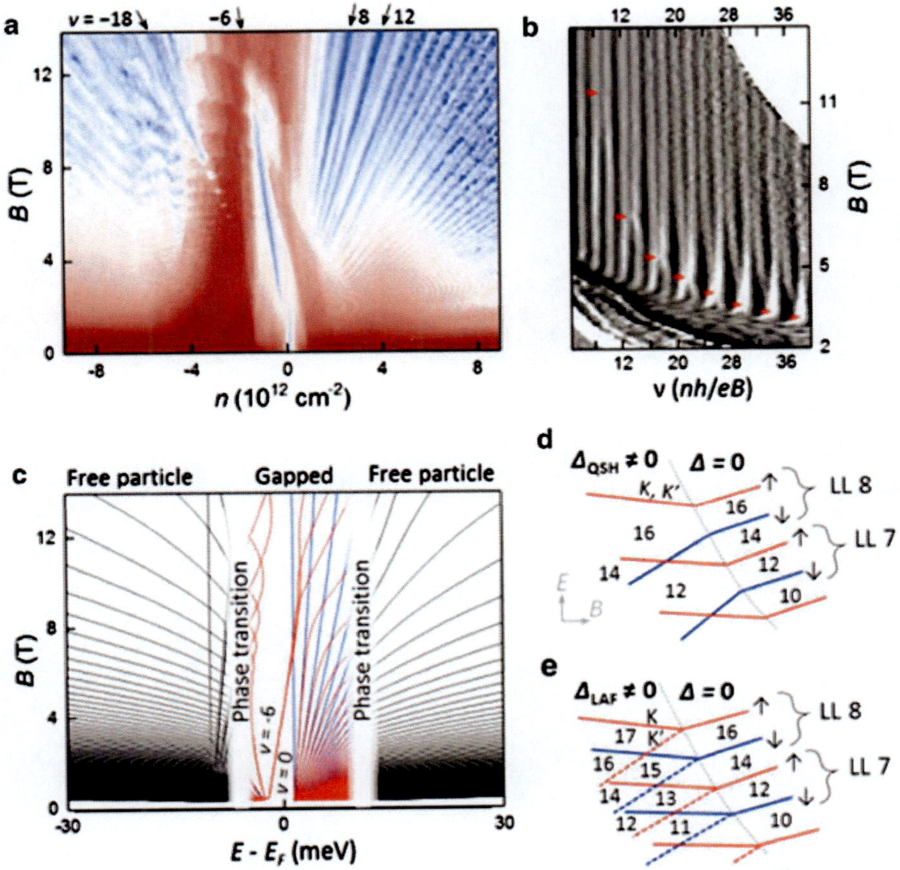

Fig. 5.5 **a** Double gated Landau fan spectrum of a 9-layer rhombohedral graphite film. **a** Map of longitudinal conductivity at 10mK at zero displacement field as a function of carrier density and magnetic field, **b** Map of differential longitudinal conductivity at 10 mK as a function of filling factors. **c** Calculated Landau level spectrum of a 9-layer rhombohedral graphite film using a gapped band structure at low energies **d**, **e** Landau level crossing for quantum spin Hall and layer antiferromagnetic order parameters, respectively. The red and blue lines denote up and down spins, respectively. The grey dotted line depicts the transition between gapped and gapless electronic states

References

1. Koshino M, McCann E (2009) Trigonal warping and Berry's phase N-pi in ABC-stacked multilayer grapheme. Phys Rev B-Condens Matter Mater Phys 80(16):1–8. https://doi.org/10.1103/PhysRevB.80.165409
2. Min H, MacDonald AH (2008) Electronic structure of multilayer graphene. Prog Theor Phys Suppl 176(176):25. https://doi.org/10.1143/PTPS.176.227
3. Slizovskiy S, McCann E, Koshino M, Fal'ko VI (2019) Films of rhombohedral graphite as two-dimensional topological semimetals, pp 1–15. Available: http://arxiv.org/abs/1905.13094
4. Shi Y et al (2020) Electronic phase separation in multilayer rhombohedral graphite. Nature 584(7820):210–214. https://doi.org/10.1038/s41586-020-2568-2

Chapter 6
Spontaneous Gap Opening in at Charge Neutrality Point of Rhombohedral Graphite Films

6.1 Introduction

Despite the theoretically predicted $E \sim p^N$ dispersion not being validated by Landau fan spectroscopy; rhombohedral graphite films still contain flat bands that are susceptible to interactions at charge neutrality point as well as a Berry phase of $N\pi$ [1, 2]. Albeit hBN encapsulated bilayer graphene devices requiring the presence of a finite displacement field to have a gap opening [3, 4], there has been variety of interaction induced spontaneous gap opening on suspended bilayer graphene systems as well as a nematic state [5–9]. The work has also been extended to suspended devices of ABC stacked tri-layer and tetra-layer systems [10, 11].

As it has been summarised on Chap. 1 in detail, the predictions on low energy bands of rhombohedral graphite films include spontaneous quantum Hall states [12] and superconductivity [13], which becomes BCS like once full tight model giving rise a quadratic low energy dispersion is considered [14]. More recently a density functional theory accompanied angular resolved photoemission spectroscopy (ARPES) experiments were carried out on rhombohedral graphite films of a 14-layer thickness [15], with the obtained data being interpreted by the presence of a layer antiferromagnetic gap, which is ferrimagnetic should the top and bottom layers be considered separately [16]. This is not surprise with the mean field calculations suggesting a layer-antiferromagnetic order parameter being ground state in bilayer graphene systems [17].

As already pointed out in the cooling curves on the previous chapter, samples ranging from 3 to 4.5 nm thickness, which corresponds to a layer number of 9–12 layers, a rise in resistance was found, in contrast to an decrease in resistance in thicker samples, due to the growing influence of surface states at low temperatures. The rise in resistance reported in higher quality 9 to 12-layer systems is due to emergence of an interaction induced transport gap at the charge neutrality point of rhombohedral graphite films. While some samples in the reported thickness range showed hysteretic behaviour as a function of gate (doping) as well as perpendicular magnetic field, in

S. Ozdemir, *Electronic Properties of Rhombohedral Graphite*, Springer Theses,
https://doi.org/10.1007/978-3-030-88307-2_6

other samples the hysteretic behaviour with respect to gate was found to be absent, a hysteresis, however, being observed as a function of in plane magnetic field.

6.2 Thermal Activation Gap Accompanied by Topological Currents

Figure 6.1a shows charge neutrality point of a 10-layer film at varying temperatures from 0.25 to 15 K with the resistance being above 10 kΩ at 0.25 K while decreasing monotonously at higher temperatures and becoming below 1 kΩ at 15 K, implying

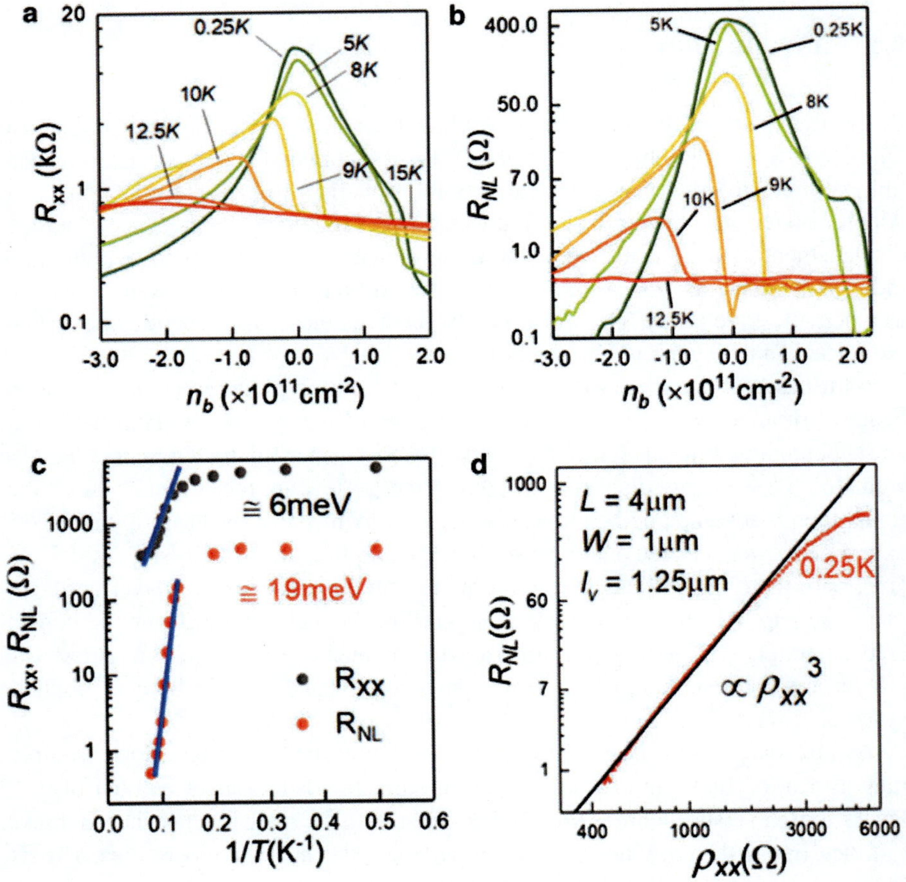

Fig. 6.1 **a** Temperature dependence of 4 probe resistance of a single gated 10-layer rhombohedral graphite film around the charge neutrality point. **b** Temperature dependence of non-local resistance emerging around charge neutrality point. **c** Arrhenius plot of peak local (grey) and non-local (red) resistance, yielding activation energies of 6 and 19 meV. **d** Non-local resistance at 0.25 K plotted against resistivity with a non-local signal fit yielding a valley decay length of 1.25 μm

a presence of thermal activation across a transport gap. Given the rich topological properties of the low energy surface Bloch bands, one may well expect topological currents with the presence of a gap, hence non-local resistance around the charge neutrality point was also checked, and indeed found as shown in Fig. 6.1b. Figure 6.1c shows an Arrhenius plot both measured peak local and non-local resistances, yielding activation energy related transport gaps of 6 meV and 19 meV, respectively. We attribute the shift of the charge neutrality point to negative carrier densities with increasing temperatures to thermally activated donor defects. The approximately factor of 3 larger non-local activation gap suggests a cubic dependence of non-local resistance to resistivity, as it would be expected from a Berry curvature induced (spin) valley Hall effect [3, 4, 18] which has indeed been found to be the case as shown in Fig. 6.1d. Fitting was carried out using the model

$$R_{NL} = \frac{1}{2}\left(\sigma_{xy}^{v}\right)^{2} \rho_{xx}^{3} \frac{W}{l_v} e^{-L/l_v} \tag{6.1}$$

with the obtained valley decay length being 1.25 μm which is comparable to high quality hBN encapsulated graphene devices. The emerging resistance accompanied by edge currents, the thermal activation behaviour, and the cubic dependence of non-local signal on resistivity all suggest presence of a finite transport gap, and zero field Hall conductance arising due to Berry curvature, which is a consequence of inversion symmetry breaking in a system possessing Berry phase on its Bloch bands.

6.3 DC Characterisation of the Gapped Resistive State

Now that we have established presence of a gapped resistive state, further characterisation is necessary to comment on the type of order parameter responsible for the gapped state. Figure 6.2a shows voltage-current characterisation at the charge neutrality point, where it can be seen that above a critical voltage drop across the sample, and/or a critical DC current driven through the sample, a negative differential resistance (NDR) is onset. We believe that, the onset of negative differential resistance is due to a voltage drop across the sample that overcomes the transport gap, where for large DC current the gap becomes smaller due to Joule heating, eventually becoming completely killed, where the I–V curve recovers to a linear slope, with smaller gradient and hence less resistance. Hence, albeit being dependent on the length of the sample at which the voltage drop is being measured, we believe the voltage drop at which the NDR onsets is related to the transport gap size. Indeed, as expected, with increasing temperature (Fig. 6.2b), the NDR onset voltage, in other words the voltage drop to overcome the band gap decreases, being completely absent at 12.5 K, suggesting a dynamic nature of the gap emerging below 12.5 K. This is consistent with the non-local signal accompanied by the gap vanishing completely at 12.5 K (Fig. 6.1b). Moreover, the voltages at which NDR onset occurs are found

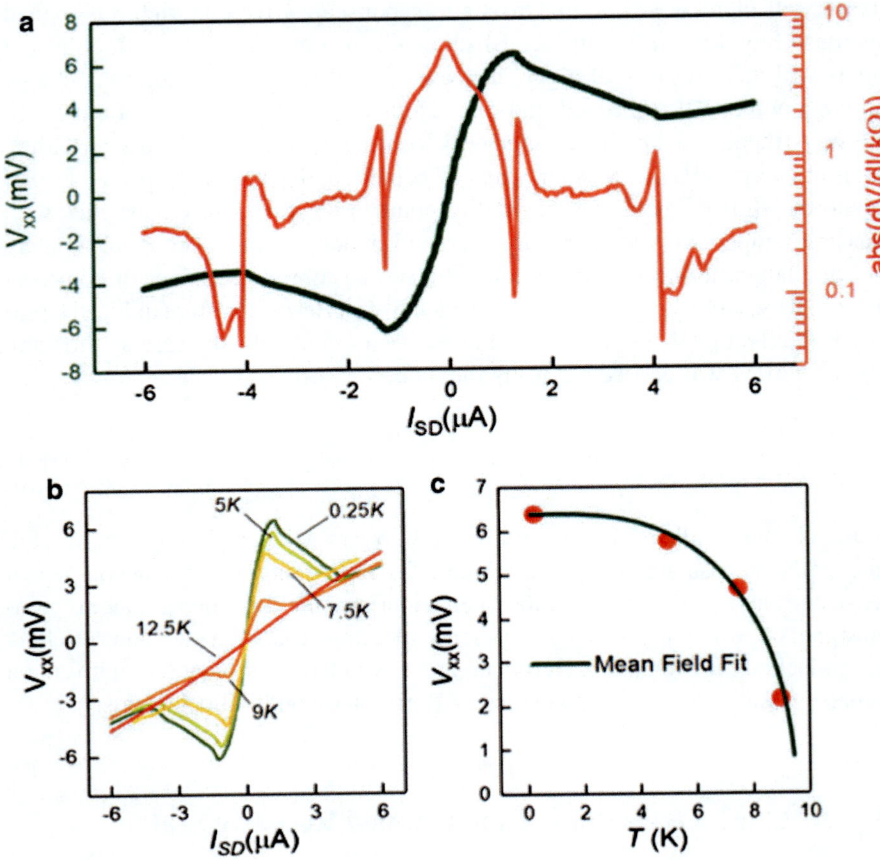

Fig. 6.2 **a** Four-probe Vxx/I curve (green) and its absolute derivative (red) with negative differential resistance regions present above a critical voltage drop and/or current across the sample at T = 0.25 K. **b** Temperature dependence of Vxx/I curves up to 12.5 K. **c** Peak Vxx values at which the negative differential resistance onsets fit by $\Delta = \Delta_0 \tanh\left(1.74\sqrt{T_C/T - 1}\right)$ mean field expression

to be remarkably well described by a mean field (BCS) fit, further suggesting the presence of spontaneous symmetry breaking in the gapped state.

6.4 Magnetic Field Dependence

Having established presence of an NDR onset, which is related to a critical voltage drop across the sample, we will now characterise the gapped resistive state as a function of magnetic field in both perpendicular and planar orientations. Figure 6.3a shows perpendicular field map (at T = 5 K where quantisation effects are absent) of differential resistance where it can be seen that the gapped resistive state (dark red

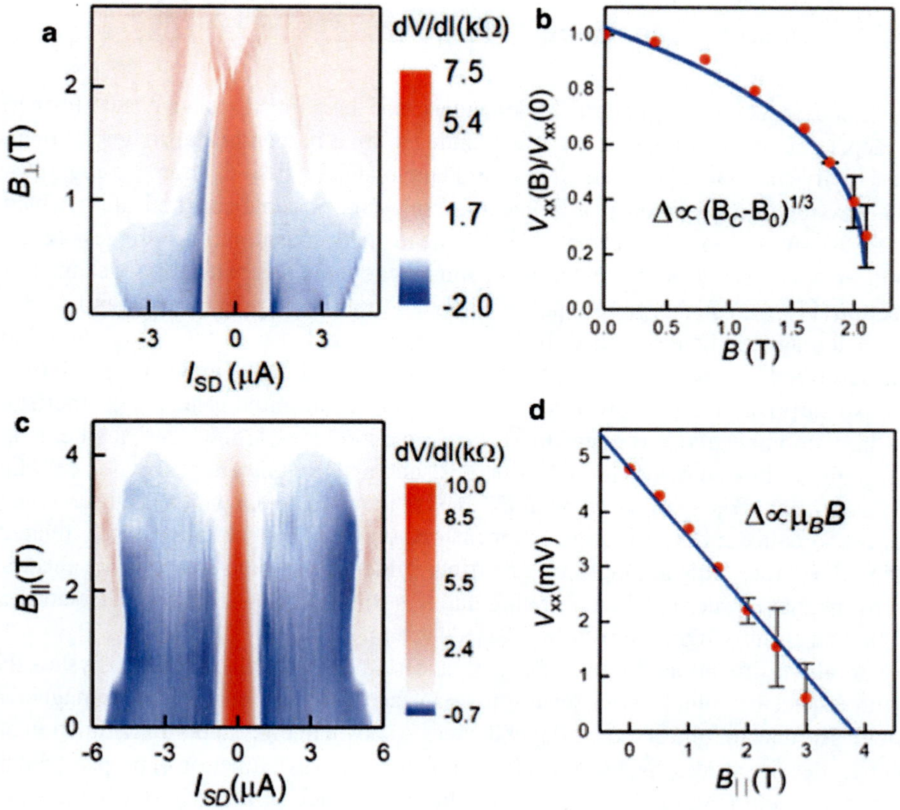

Fig. 6.3 Destruction of the gapped resistive state by magnetic field of perpendicular and in-plane orientations. **a** Map of differential voltage-current for a DC current sweep range of ±6 μA and a stepped magnetic field range of 0–2.5 T at T = 5 K. **b** Normalised gap size related NDR onset critical voltage drop across the sample at various magnetic field fields at T = 5 K (red) fitted with the mean field magnetisation scaling relation $(B_C - B)^{1/3}$ (blue) where B_C is the critical field at which gap vanishes. **c** Map of differential voltage-current for a DC current sweep range of ±6 μA and a stepped in plane magnetic field range of 0–4.5 T at T = 1.6 K. **d** Gap size related NDR onset critical voltage at varying in plane magnetic fields (red) fitted with a linear magnetic field dependence (blue) expected from the Zeeman energy enhancement by the in-plane magnetic field

regions) and the NDR (blue regions) with the gap becoming completely killed at 2 T. Figure 6.3b shows the normalised NDR onset voltage as a function of magnetic field, with the blue fit being a $(B_C - B)^{1/3}$ scaling law. Such a scaling is well known within the mean-field description of systems possessing a finite magnetisation [19], hence directly suggests magnetisation property of the gapped phase, which may well be an orbital magnetisation arising from finite Berry curvature. The NDR onset voltage has been found to have a more linear dependence on the applied magnetic field when its applied parallel to the sample (Fig. 6.3c, d).

6.5 Hysteretic Behaviour

To characterise the magnetic field dependence of the gapped resistive state further, independent of the quantisation effects induced by a perpendicular magnetic field, resistivity was studied as a function of in plane magnetic field as shown in Fig. 6.4a. The hysteretic behaviour at a first glance suggests possible magnetisation which is not a surprise given the fact that the mean field considerations suggest layer-antiferromagnetic order to be dominant and magnetic field dependent I–V characterisation (Fig. 6.3) on this gapped phase. Figure 6.4b shows temperature dependence of critical magnetic field obtained from the magnetic field sweeps which was found to be described by a power law, depicted as an inset. Remarkably, and at the same time not so surprisingly, this power law, or scaling has been found to be a characteristic of large-moment antiferromagnetic (LAFM) and hidden order phases that emerge in strongly correlated heavy fermion systems such as URu_2Si_2 and $CsFeCl_3$ [20, 21]. The critical exponent of the fit obtained, α, was found to be 0.4 which is closer to one of the so-called hidden order phases measured on URu_2Si_2, which is closely related to LAFM order. This finding is also a further evidence that points us towards an antiferromagnetic order parameter arising due to strong electron–electron correlations, albeit an entirely light atom carbon-based system.

As already mentioned at the introduction, certain samples also exhibited hysteresis with respect to applied carrier density at the surface, as well as perpendicular magnetic field. The data for one of these 9-layer devices is shown in Fig. 6.5 both as a function of carrier density at zero displacement field (Fig. 6.5a) and as a function of perpendicular magnetic field (Fig. 6.5b). We attribute the occurrence of these hysteresis events, with multiple abrupt resistivity changes as the phase transition is taking place, to the presence of multiple mesoscopic domains of this gapped state. This idea could be

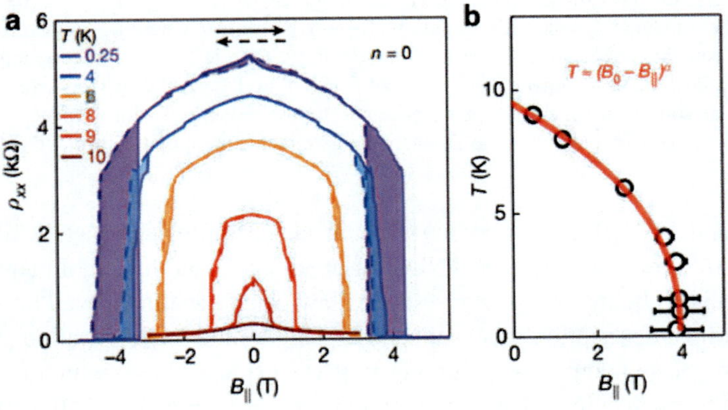

Fig. 6.4 **a** Hysteresis in resistivity as the spontaneous gap is killed while a planar magnetic field is swept at the charge neutrality point. **b** Extracted critical magnetic field values at temperatures up to 9 K, obeying the power law shown in the inset with red line being a fit where the coefficient α is set to 0.4

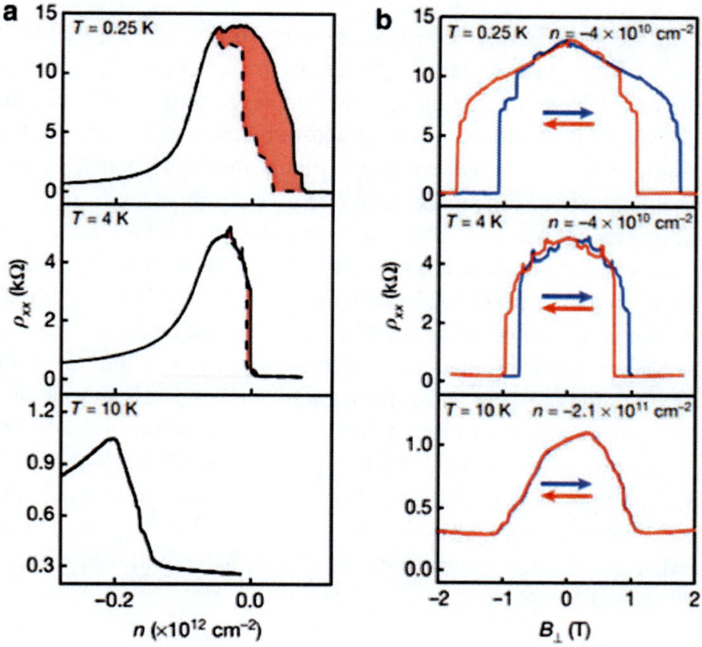

Fig. 6.5 **a** Temperature dependence of hysteresis in resistivity at the charge neutrality point **a** as carrier density is tuned at D = 0 on a 9-layer device **b** as the perpendicular magnetic field is swept

verified using advanced scanning probe microscopy techniques which could also be helpful on commenting on the nature of the domains.

6.6 Summary

We have identified emergence of a spontaneous interaction induced transport gap with peculiar symmetry properties on the quadratically dispersing flat bands of rhombohedral graphite films between 9 and 12-layer thickness. It has been found that this gap exhibits thermal activation behaviour yielding an Arrhenius gap size of ~6 meV. Since the emergence of the gap breaks the A/B sublattice inversion symmetry, it has been found to be accompanied by a non-local signal. This non-local signal was found to be well described by a valley Hall effect model, where a zero magnetic field Hall conductance arising due to Berry curvature is considered.

I–V characterisation at the charge neutrality point was found to yield a negative differential resistance onset, most likely due to the potential gain of each electron overcoming the transport gap, with the gap simultaneously being supressed due to Joule heating. The voltage drop across the Vxx contacts at which NDR was onset is found to be well described by a mean field gap expression, which is consistent with the emergence of the gap due to a spontaneous symmetry breaking. Magnetic field dependence of the I–V curves showed suppression of the gap as well as the NDR onset voltage in both perpendicular and planar orientations of the field. Perpendicular field dependence of NDR onset voltage was found to yield a cubic dependence to the magnetic field as it would be expected within mean field framework from a gap possessing finite magnetisation whereas the planar magnetic field was found to be linear, as would be expected from Zeeman energy enhancement.

Remarkably, a hysteresis was found to be present when planar magnetic field characterisation was carried out, which is not a surprise given most likely spin dependent nature of the responsible order parameter for the gap, and hence a probable spin ordering. Moreover, when the critical field obtained at each temperature was plotted it was found to be remarkably well described by a power law that has been a characteristic of both large moment antiferromagnetic order as well as a hidden order in heavy fermion systems, demonstrating rhombohedral graphite films as an alternative platform for strongly correlated electron physics. Finally, a hysteresis as well as abrupt jumps in resistance was also found as both the carrier density and perpendicular magnetic field was tuned on some but not all samples, suggesting presence of multiple mesoscopic domains on certain samples, which could be a very interesting scanning probe microscopy study.

References

1. Koshino M, McCann E (2009) Trigonal warping and Berry's phase N-pi in ABC-stacked multilayer grapheme. Phys Rev B Condens Matter Mater Phys 80(16):1–8. https://doi.org/10.1103/PhysRevB.80.165409
2. Slizovskiy S, McCann E, Koshino M, Fal'ko VI (2019) Films of rhombohedral graphite as two-dimensional topological semimetals, pp 1–15. [Online]. Available: http://arxiv.org/abs/1905.13094
3. Shimazaki Y, Yamamoto M, Borzenets IV, Watanabe K, Taniguchi T, Tarucha S (2015) Generation and detection of pure valley current by electrically induced Berry curvature in bilayer graphene. Nat Phys 11(12):1032–1036. https://doi.org/10.1038/nphys3551
4. Sui M et al (2015) Gate-tunable topological valley transport in bilayer graphene. Nat Phys 11(12):1027–1031. https://doi.org/10.1038/nphys3485
5. Martin J, Feldman BE, Weitz RT, Allen MT, Yacoby A (2010) Local compressibility measurements of correlated states in suspended bilayer graphene. Phys Rev Lett 105(25):256806. https://doi.org/10.1103/PhysRevLett.105.256806
6. Weitz RT, Allen MT, Feldman BE, Martin J, Yacoby A (2010) Broken-symmetry states in doubly gated suspended bilayer graphene. Science 330(6005):812–816. https://doi.org/10.1126/science.1194988
7. Mayorov AS et al (2011) Interaction-driven spectrum reconstruction in bilayer graphene. Science 333(6044):860–863. https://doi.org/10.1126/science.1208683
8. Freitag F, Trbovic J, Weiss M, Schönenberger C (2012) Spontaneously gapped ground state in suspended bilayer graphene. Phys Rev Lett 108(7):076602. https://doi.org/10.1103/PhysRevLett.108.076602
9. Velasco J et al (2012) Transport spectroscopy of symmetry-broken insulating states in bilayer graphene. Nat Nanotechnol 7(3):156–160. https://doi.org/10.1038/nnano.2011.251
10. Lee Y et al (2014) Competition between spontaneous symmetry breaking and single-particle gaps in trilayer graphene. Nat Commun 5:5656. https://doi.org/10.1038/ncomms6656

11. Myhro K et al (2018) Large tunable intrinsic gap in rhombohedral-stacked tetralayer graphene at half filling. 2D Mater 5(4). https://doi.org/10.1088/2053-1583/aad2f2
12. Zhang F, Jung J, Fiete GA, Niu Q, MacDonald AH (2011) Spontaneous quantum hall states in chirally stacked few-layer graphene systems. Phys Rev Lett 106(15):1–4. https://doi.org/10.1103/PhysRevLett.106.156801
13. Kopnin NB, Heikkilä TT, Volovik GE (2011) High-temperature surface superconductivity in topological flat-band systems. Phys Rev B Condens Matter Mater Phys 83(22):1–4. https://doi.org/10.1103/PhysRevB.83.220503
14. Kopnin NB, Ijäs M, Harju A, Heikkilä TT (2013) High-temperature surface superconductivity in rhombohedral graphite. Phys Rev B Condens Matter Mater Phys 87(14):1–4. https://doi.org/10.1103/PhysRevB.87.140503
15. Henck H et al (2018) Flat electronic bands in long sequences of rhombohedral-stacked graphene. Phys Rev B 97(24):1–6. https://doi.org/10.1103/PhysRevB.97.245421
16. Pamuk B, Baima J, Mauri F, Calandra M (2017) Magnetic gap opening in rhombohedral-stacked multilayer graphene from first principles. Phys Rev B 95(7). https://doi.org/10.1103/PhysRevB.95.075422
17. Kharitonov M (2012) Canted antiferromagnetic phase of the $\nu=0$ quantum Hall state in bilayer graphene. Phys Rev Lett 109(4):1–5. https://doi.org/10.1103/PhysRevLett.109.046803
18. Gorbachev RV et al (2014) Detecting topological currents in graphene superlattices. Science 346(6208):448–451. https://doi.org/10.1126/science.1254966
19. Hook JR, Hall HE (1990) Solid state physics. Wiley
20. Ran S et al (2017) Phase diagram of $URu_{2-x}Fe_xSi_2$ in high magnetic fields. Proc Natl Acad Sci U S A 114(37):9826–9831. https://doi.org/10.1073/pnas.1710192114
21. Kurita N, Tanaka H (2016) Magnetic-field- and pressure-induced quantum phase transition in $CsFeCl_3$ proved via magnetization measurements. Phys Rev B 94(10):1–7. https://doi.org/10.1103/PhysRevB.94.104409

Chapter 7
Displacement Field Induced Band Gap Opening in Rhombohedral Graphite Films

7.1 Introduction

A displacement field induced band gap opening because of inversion (sublattice) symmetry breaking has been demonstrated in both bilayer graphene [1–4] and ABC stacked tri-layer systems [5]. Given the closely related nature of low energy dispersion in N-layer rhombohedral graphite films, when considering only the primary hopping parameters, a displacement field induced band gap opening has also been predicted [6], above a critical field which depends on the number of layers as outlined already in Chap. 1, Sect. 1.14. Moreover, given the calculated sensitivity of such a displacement field induced gap to stacking and stacking faults in tetra-layer systems [7], displacement field induced gap opening may indeed be a signature of pristine rhombohedral stacking on graphite films even on films up to 50 layers thickness.

7.2 Displacement Field Induced Resistivity Increase

Figure 7.1 shows displacement field-carrier density maps on rhombohedral graphite films ranging from 9 to 50 layers thickness which were identified through Raman spectroscopy prior to transport characterisation. The spontaneous gap on 9-layer system at zero displacement field rapidly closes, considered in detail in Chap. 6, while a new gap emerges above a critical displacement field. Qualitative agreement to gap opening prediction of Koshino [6] can immediately be seen which is in principle a result of the theoretically expected low energy two-band structure of rhombohedral graphite. The increase in resistivity, evident above a critical displacement field for all pristine rhombohedral graphite systems, high resistivity points which corresponds to the white to red regions of the contour plot, is because of band gap opening and indeed obeys an Arrhenius dependence as expected from generation of such a band gap (see Sect. 7.4).

S. Ozdemir, *Electronic Properties of Rhombohedral Graphite*, Springer Theses,
https://doi.org/10.1007/978-3-030-88307-2_7

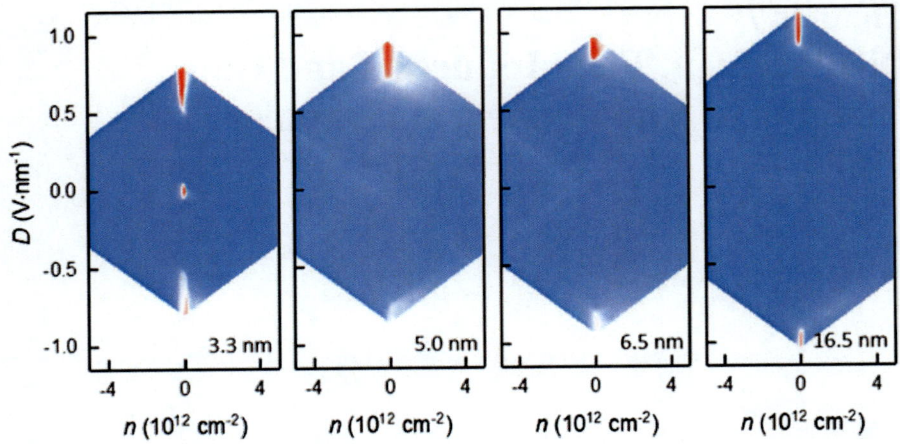

Fig. 7.1 Resistivity, ρ_{xx} (n, D) maps of rhombohedral graphite films with thicknesses ranging from 3.3 to 16.5 nm where the colour scale from light blue to red corresponds to 10 Ω–25 kΩ, 7 kΩ, 5 kΩ, 2 kΩ on each map respectively. The maps were obtained at temperatures of around T = 0.25 K

Figure 7.2a shows line cuts from the displacement field maps where the resistivity is normalised. The extracted values of Dc for varying sample thicknesses are shown in Fig. 7.2b, with the red fit being a self-consistent model obtained for the low energy band structure which is obtained through firstly an estimation of the screening carrier density at top and bottom surfaces as [8]

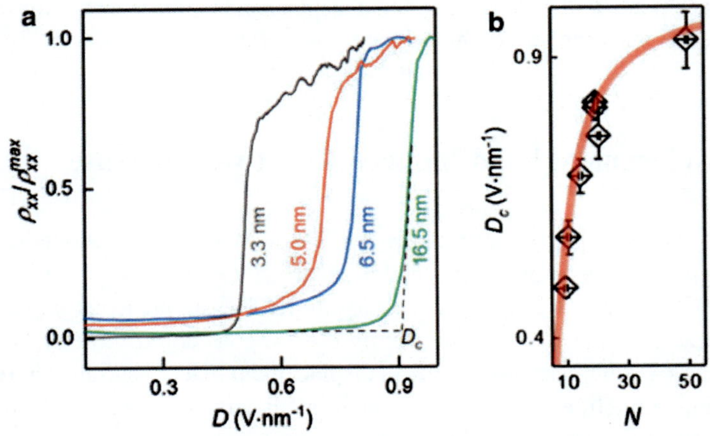

Fig. 7.2 **a** Line cuts of the displacement field maps with respect to normalised resistivity where the dashed lines depict extraction of critical field, D_c for each film. **b** Critical field, D_c plotted as a function of layer number, with the red fit being a numerical calculation carried out using the full tight-binding model

$$n_i^{scr} = 2\int_{BZ} \frac{d^2k}{(2\pi)^2} \sum_{i=1}^{2N} \left[\left(\left| \psi_{Ai}^l(\mathbf{k}) \right|^2 + \left| \psi_{Bi}^l(\mathbf{k}) \right|^2 \right) f(\varepsilon_l - E_F) - 1 \right] \tag{7.1}$$

where the wavefunction $\psi_{A/Bi}^l(\mathbf{k})$ for each band on each sublattice is given by the full SWCM model implied by the Landau fan spectroscopy with index i corresponding to the particular layer, and l corresponds to the particular orbital considered. The constant -1 takes care of the charges that develop to neutralise the sample at the Fermi energy. It has to be reminded here that, the wavefunction amplitudes for low energy bands are extremely localized on the top and bottom surfaces, hence they are the ones that give rise to almost all of the charges on top and bottom surfaces as the bulk bands are more localised on the bulk rather than the surfaces. As well as being dependent on the momentum k within the Brillouin zone, the wavefunction amplitudes in Eq. (7.1) are coupled to the band gap generated between the conduction and valance bands, which we can denote as Δ^*. Within the self-consistent Hartree model, Δ^* is related to the externally applied displacement field D as

$$D = \frac{-e(n_t - n_b)}{2\varepsilon_0} = \frac{-e\left(n_t^{scr}(\Delta^\sim) - n_b^{scr}(\Delta^\sim)\right)}{2\varepsilon_0} + \frac{\varepsilon_r \Delta^\sim}{d} \tag{7.2}$$

where ε_r is the permittivity of rhombohedral graphite thin film assumed to be same as that of graphene.

As well as the change in the low energy dispersion property for increasing number of layers another important factor that had to be considered to calculate theoretically the values of D_{crit} above which a transport band gap, Δ, emerges is the band overlap, with size of ${}^{2\gamma_4\gamma_1}/_{\gamma_0}$. The transport gap therefore is related to the band gap as

$$\Delta = \Delta^* - \frac{2\gamma_4\gamma_1}{\gamma_0} \tag{7.3}$$

which is depicted on Fig. 7.3a on a band structure schematic. Figure 7.3b shows the relation of both gaps to the applied displacement field for a 9-layer thick film, and it can be seen that as a result of the finite band overlap, the transport gap emerges at a much later critical displacement compared to the masked band gap. Albeit the masking effect being significant, it must be noted that its (for thickness of devices studied) layer number independent. It is the changes in number of screening charges that lead to the layer dependent difference on the displacement field maps.

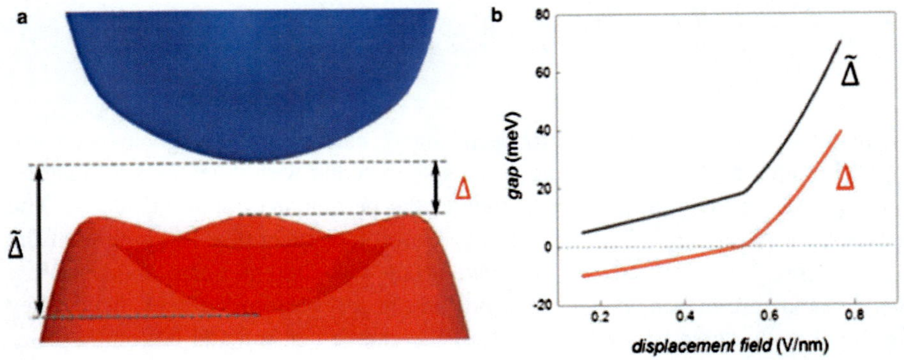

Fig. 7.3 **a** Gapped low energy band structure of a 9-layer rhombohedral graphite film with masked band gap, and the transport band gap illustrated. **b** A comparison of the dependence of the size of the masked band gap, and transport band gap on the externally applied displacement field

7.3 Absence of a Gap Opening for Graphite Films of Mixed Stacking

Now that we have established both experimentally and theoretically that there is a finite transport gap opening in rhombohedral graphite films, the next major thing to discuss is whether one can use the presence of such a displacement field induced gap opening to the stacking faults [8]. The literature on electronic properties of stacking faults was extensively discussed on Chap. 1, Sect. 1.17. For our electronic investigations 3 types of stacking faults were considered which are namely twinned boundary, Bernal hexagonal and surface Bernal hexagonal where the structures could be illustrated as ABCABCBACBA, ABCBCABC, ABCABCABCABA where each stack could be extended to many number of layers but it should be noted that for the third stack ABA region is where the stacking terminates. The three types of stacking faults are illustrated pictorially on Fig. 7.4a, d, g. Figure 7.4b, c show low energy band structure of an 18-layer thick system with a twinned boundary stacking fault with a displacement field of 0.9 to −0.9 V/nm and absence of a transport band gap is clearly visible. Figure 7.4e, f, h, i show a rhombohedral stack with a Bernal hexagonal stacking fault and a surface Bernal hexagonal stacking fault respectively at displacement field of 0.9 to −0.9 V/nm. In both cases a transport gap has been found to be absent. Low energy band structures were calculated in the vicinity of K (K′) point through appropriate orbital linkage relations for each type of stacking fault using the full tight binding model implied by Landau fan spectroscopy.

As discussed in Chap. 1, Sect. 1.18, Raman spectroscopy, and the double resonant 2D peak is a well-established method of characterising stacking order in graphite films prior to electronic characterisation. The larger full width at half maximum (FWHM) of the 2D peak, as well as a shoulder at the left end of the 2D peak has been calculated and found to be signatures of pristine rhombohedral graphite stacks. Typically, once Raman mapping of a rhombohedral graphite flake is carried out, it

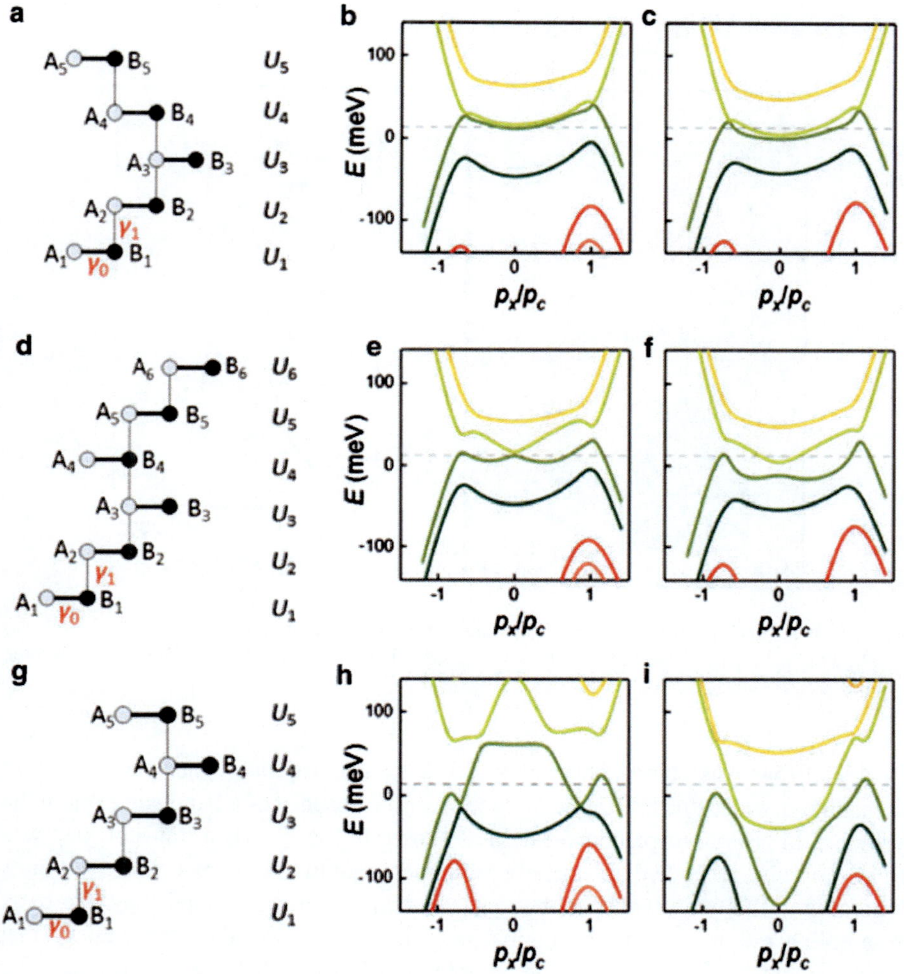

Fig. 7.4 Stacking faults on rhombohedral graphite films **a** schematic of a twinned boundary type stacking fault, and the low energy band structure for **b** positive and **c** negative displacement field of 0.9 V/nm for an 18-layer graphite film. **d** Schematic of a Bernal hexagonal stacking fault, and the low energy band structure for **e** positive and **f** negative displacement field of 0.9 V/nm for an 18-layer graphite film. **g** Schematic of a surface Bernal hexagonal type stacking fault and the low energy band structure for **h** positive and **i** negative displacement field of 0.9 V/nm

is possible to see a range of FWHM of the 2D peak as shown in Fig. 7.5a, with maximum FWHM regions should correspond to regions of pristine rhombohedral graphite (#c) whereas the minimum FWHM regions being pristine hexagonal Bernal stacking (#a) and finally any intermediate FWHM regions (#b) corresponding to the graphite regions with stacking faults. As well as the map, sample 2D peaks are plotted on Fig. 7.5b, with the left shoulder pointed by the black arrow disappearing for intermediate FWHM region, in agreement with first principles calculations.

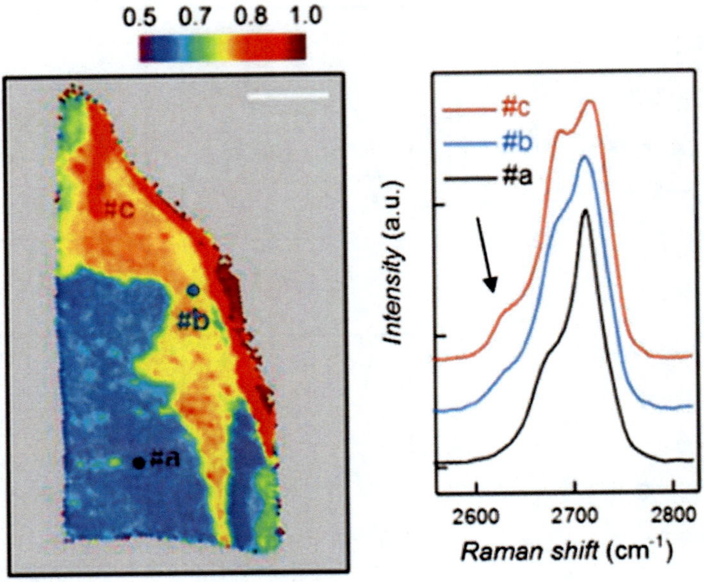

Fig. 7.5 **a** Raman spectrum 2D peak full width half maximum map of a graphite flake 6.5 nm graphite flake with varying FWMH values at regions a, b, c. **b** Raman spectra 2D peak on the three regions a, b, c

A Hall bar was etched onto the region #b and a top-back gate map of it was measured in order to confirm the prediction of the absence of a band gap even in the presence of a large displacement field as shown in Fig. 7.6. If a transport gap was to be emerging at large displacement field, there would have been a resistive points emerging at the two corners of the map, which has not been found to be the case, resembling rather a Bernal hexagonal graphite system. We therefore believe that displacement field characterisation of graphite films enables one to verify pristine rhombohedral stacking in films of up to and including 50 layers.

7.4 Non-local Signal Due to Displacement Field Induced Gap Opening

Now that we have established the absence of stacking faults in our rhombohedral graphite films, and displacement field generated band gap as a transport signature of this, we will be returning to the gap generated at large displacement fields, as a particular example, we will be considering the 9-layer device for which data is shown on Fig. 7.7. Figure 7.7a is an Arrhenius plot of resistance at displacement fields ranging from 0.57 to 0.76 V/nm and a linear fit at each displacement field can be seen, and yields an activation energy which are plotted on the inset with the red fit being the gap size estimated with the self-consistent model outlined. Figure 7.7b

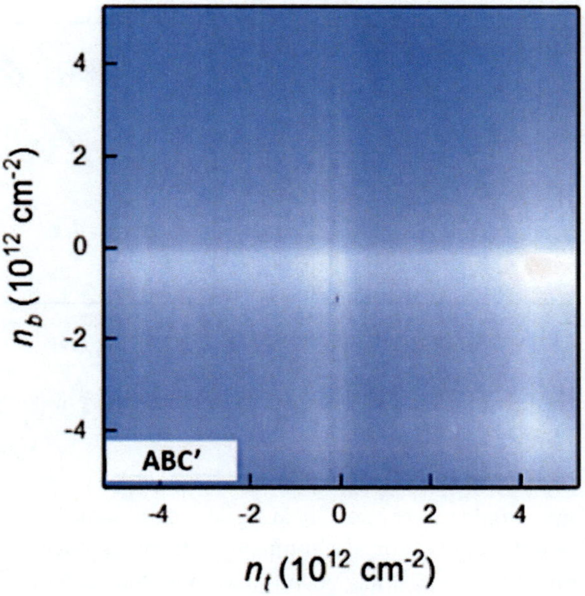

Fig. 7.6 Top gate back gate map of a 6.5 nm graphite film device with an inferred mixed stacking from Raman spectroscopy measured at T = 1.6 K with the colour scale from blue to white to red corresponding to 20–70 Ω

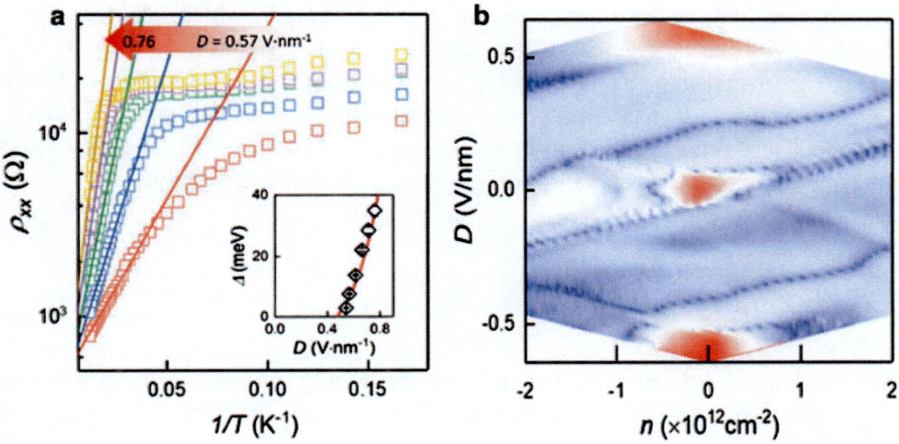

Fig. 7.7 **a** Arrhenius plot enabling the displacement field generated band gap extraction on a 9-layer rhombohedral graphite film with the extracted gap size plotted in the inset. **b** Non-local resistance measured at T = 1.6 K as a function of carrier density and displacement field with the colour scale blue to red corresponding to 0.1 Ω–10 kΩ

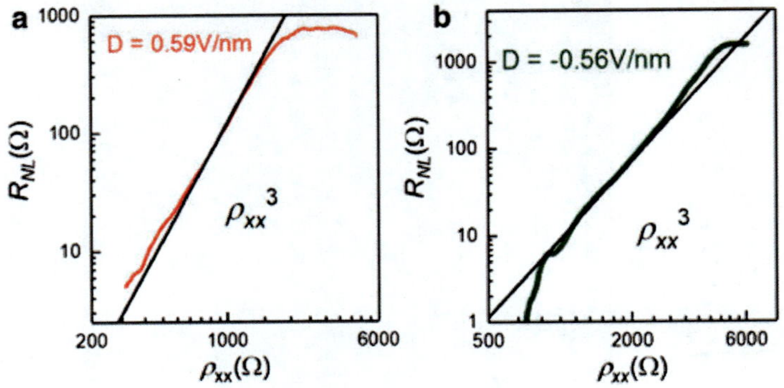

Fig. 7.8 **a** Non-local signal plotted against resistivity at finite values of positive and negative displacement fields measured at T = 0.25 K

shows a displacement field map of non-local signal on the same 9-layer device and it can be seen that non-local resistance accompanied the emergence of a transport gap at large displacement fields, as it has been found to be the case in bilayer graphene [9, 10] and been also found the spontaneous gapped state considered in Chap. 6.

The emerging non-local signal was plotted against resistivity to see if a cubic dependence is present as expected from a Berry curvature induced valley Hall effect arising due to sublattice (inversion) symmetry breaking by the finite displacement field. Indeed, due to the Berry phase of Nπ it possesses, which becomes Berry curvature with the emergence of the gap, the non-local signal does obey a cubic dependence to resistivity as shown in Fig. 7.8 for either polarity of displacement field. Hence, we believe that the observed non-local signal emerging due to electrically generated Berry curvature is a further confirmation of rhombohedral graphite films behaving as an N-layer generalisation of graphene in low energy limit.

7.5 Summary

We have demonstrated displacement field induced band gap opening in rhombohedral graphite films of up to 50 layers thickness. We have shown that there is a strong layer number dependence on critical displacement field above which measured transport gap emerges. The layer number dependence is due to change in low energy dispersion with increasing number of layers. Moreover, we find that the induced band gap is initially masked, due to a significant band overlap caused by the hopping parameter γ_4 and this masking effect is layer independent and its relation to observed transport gap has been shown. Agreement was obtained with the experimental data through numerical calculations considering of the screening effect due to charges arising from wavefunctions localised on each surface. Stacking faults of twinned, hexagonal Bernal and surface hexagonal Bernal types were considered theoretically and found to show absence of a transport gap. To check for the accuracy of the prediction, and intermediate 2D peak FWHM regions was selected on a Raman mapped 6.5 nm graphite film with mixed stacking, with the device showing indeed absence of a band gap opening at large displacement fields. Hence, we verified displacement field transport experiments as an additional tool to verify pristine rhombohedral stacking on graphite films. Finally, the transport gap emerging in large displacement fields was found to be accompanied by a non-local signal, arising due to Berry curvature generated because of sublattice symmetry breaking and proving rhombohedral graphite as an N-layer generalisation of graphene in low energy limit. The non-local signal was found to exhibit a cubic dependence to resistivity, which is now expected as the transport signature of a valley Hall effect in the absence of spin orbit coupling effects.

References

1. Castro EV et al (2007) Biased bilayer graphene: semiconductor with a gap tunable by the electric field effect. Phys Rev Lett 99(21):8–11. https://doi.org/10.1103/PhysRevLett.99.216802
2. Oostinga JB, Heersche HB, Liu X, Morpurgo AF, Vandersypen LMK (2008) Gate-induced insulating state in bilayer graphene devices. Nat Mater 7(2):151–157. https://doi.org/10.1038/nmat2082
3. Zhang Y et al (2009) Direct observation of a widely tunable bandgap in bilayer graphene. Nature 459(7248):820–823. https://doi.org/10.1038/nature08105
4. Mak KF, Lui CH, Shan J, Heinz TF (2009) Observation of an electric-field-induced band gap in bilayer graphene by infrared spectroscopy. Phys Rev Lett 102(25):100–103. https://doi.org/10.1103/PhysRevLett.102.256405
5. Lui CH, Li Z, Mak KF, Cappelluti E, Heinz TF (2011) Observation of an electrically tunable band gap in trilayer graphene. Nat Phys 7(12):944–947. https://doi.org/10.1038/nphys2102
6. Koshino M (2010) Interlayer screening effect in graphene multilayers with ABA and ABC stacking. Phys Rev B Condens Matter Mater Phys 81(12):1–7. https://doi.org/10.1103/PhysRevB.81.125304
7. Aoki M, Amawashi H (2007) Dependence of band structures on stacking and field in layered graphene. Solid State Commun 142(3):123–127. https://doi.org/10.1016/j.ssc.2007.02.013
8. Shi Y et al (2020) Electronic phase separation in multilayer rhombohedral graphite. Nature 584(7820):210–214. https://doi.org/10.1038/s41586-020-2568-2
9. Shimazaki Y, Yamamoto M, Borzenets IV, Watanabe K, Taniguchi T, Tarucha S (2015) Generation and detection of pure valley current by electrically induced Berry curvature in bilayer graphene. Nat Phys 11(12):1032–1036. https://doi.org/10.1038/nphys3551
10. Sui M et al (2015) Gate-tunable topological valley transport in bilayer graphene. Nat Phys 11(12):1027–1031. https://doi.org/10.1038/nphys3485

Chapter 8
Three Dimensional Quantum Interference of Bulk Electrons

8.1 Introduction

Albeit low energy topological surface bands attracting the main interest, the gapped bulk sub-bands of rhombohedral graphite films are also peculiar. They possess a linear dispersion for in plane direction and a quantised momentum in k_z direction [1] described as

$$\varepsilon_n = \pm\sqrt{(vp)^2 + \gamma_1^2 + 2\gamma_1 vp\cos(k_z^d d)} \tag{8.1}$$

with values of k_z being obtained at each value of in plane momentum p from solutions of

$$vp\sin((N+1)k_z d) + \gamma_1 \sin(Nk_z d) = 0 \tag{8.2}$$

where (N − 1) real k_z solutions exist for Eq. (8.2) [1], with in plane momentum dependent k_z values describing the gapped bulk bands.

Another important point of the study of electron transport in bulk bands is previously limited studied aspect of topological nodal line semimetals [2–4] as well as their thin films in contrast to Weyl and Dirac semimetals. Here, unlike the earlier chapters focussing on two dimensional surfaces which is very much a generalisation of graphene physics, we report on bulk electrons in the high quality in thin film devices of this nodal line system and quantum interference phenomena that reflects the bulk topology. As the dispersion relation in (8.1) and the equality in Eq. (8.2) imply that for any in plane motion of massive Dirac fermions in bulk sub-bands one will also get a change in the k_z, hence leading to a 3D motion albeit a quasi 2D thin film sample. This, despite the 3D chiral anomaly physics being ruled out leads to 3D quantum interference effects [5] that reflect topological properties of massive Dirac fermions in diffusive limit as in the case of Weyl and Dirac semimetals [6–9].

S. Ozdemir, *Electronic Properties of Rhombohedral Graphite*, Springer Theses,
https://doi.org/10.1007/978-3-030-88307-2_8

8.2 Magnetic Field Rotation of a 9-Layer Film

The distinguishing feature of the gapped bulk charges in rhombohedral graphite thin films from the ones of their surface counterparts would be their three-dimensional transport. Hence, to look for any signatures of three-dimensional transport, especially that of quantum interference effects, magnetoresistance of the samples was studied with varying magnetic field orientations. The main signature of weak (anti)localization of charges arising due to quasiparticles undergoing 3D random walk, that is different from their 2D counterparts is their field orientation independent dephasing [10], unlike 2D weak (anti)localization for which magnetic field signature is only present for perpendicular field [11].

Figure 8.1a shows magnetic field rotation on a 9-layer system. A negative magnetoresistance (NMR) emerges as the magnetic field is rotated from out of plane 90° orientation to in plane 0° orientation. Moreover, this NMR is found to be supressed when conduction is surface dominated as the two surfaces of the film are doped as shown in Fig. 8.1b, hence suggesting its bulk origins. The temperature dependence of the NMR at zero doping for current aligned magnetic field are shown in Fig. 8.1c and, albeit being strongly supressed with increasing temperature, the observed NMR remains present with a quadratic field dependence resembling chiral anomaly until room temperature. Given the quasi-2D nature of the bulk nodal line system being studied, one can rule out chiral anomaly. Although a more exotic explanation such as Berry curvature would be tempting, it is debatable how different the proposed Berry curvature effect on topological insulators is to chiral anomaly of massive quasiparticles.

Another, and the more important similarity of the 9–10-layer samples to Weyl/Dirac semimetals is the weak field magnetoconductance below 20 K as shown in Fig. 8.1d. It can be seen that, despite presence of a quadratic positive magnetoconductance there is a counteracting negative magnetoconductance at small fields, which has a square root of B dependence as expected from a 3D weak antilocalization effect (WAL). Hence the magnetoconductance was found to be described with the equation $\Delta\sigma = \alpha B^{1/2} + \beta B^2$. This square root of magnetic field dependent negative magnetoconductance, described by coefficient α, is supressed completely at 20 K being consistent with low temperature requirement for the manifestation of this quantum interference effect. Given that, the system we are studying is entirely carbon based, it is very unlikely that the 3D WAL behaviour would manifest itself because of spin orbit coupling. This leaves the Berry phase of the massive bulk Bloch band charges and their random walk in three dimensions as the only alternative, hence therefore reflecting topological properties of the bulk of this nodal line semimetal thin film.

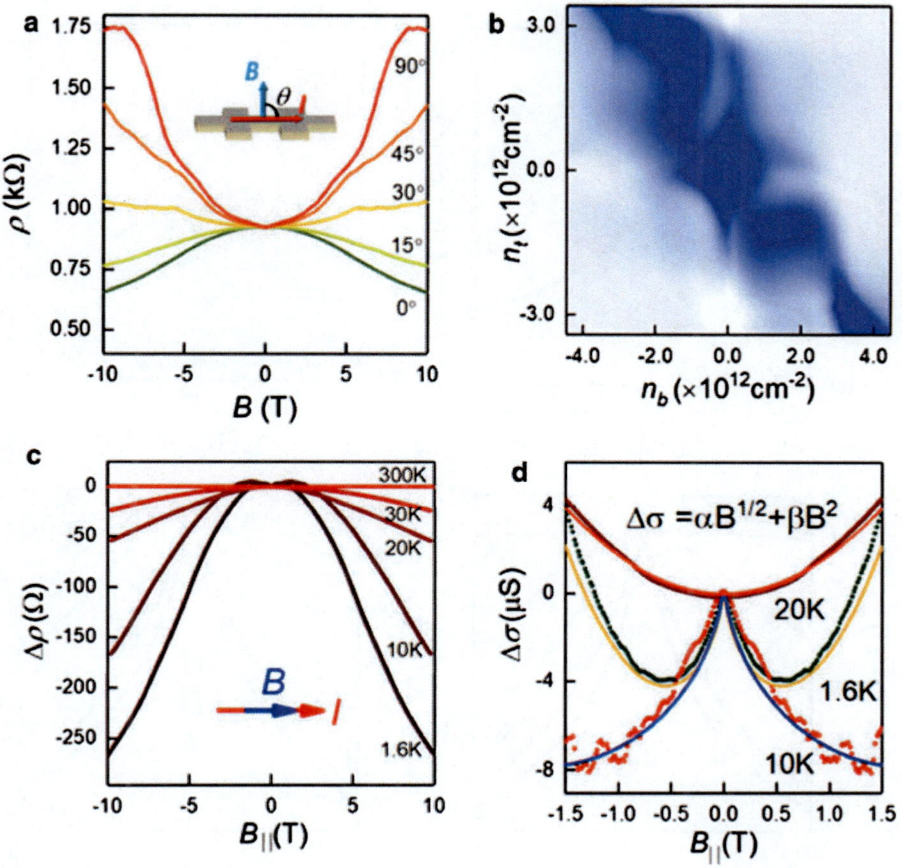

Fig. 8.1 Quadratic negative magnetoresistance accompanied by weak antilocalization on a 9-layer rhombohedral graphite film. **a** Out of plane (90°) to in plane (0° current aligned) magnetic field rotation at 1.6 K. **b** Top gate back gate map of change in resistance from 0 to 10 T magnetic field at 1.6 K with the colour scale blue to white corresponding to −250 to 0 Ω. **c** Temperature dependence of negative magnetoresistance due to current aligned magnetic field **d** Magnetoconductance as a function of weak current aligned field (±1.5 T) at 1.6 K (green), 10 K (red) and 20 K (brown) as well as respective fits of the equation $\sigma = \alpha B^{1/2} - \beta B^2$ where α is −9, −9, 0 μS/T$^{1/2}$ and β is 5.5, 1, 1.75 μS/T^2 for yellow (1.6 K) blue (10 K) and orange (20 K) fits respectively

8.3 Magnetic Field Rotation of a 14-Layer Film

The in plane magnetic field rotation was also carried out on films of greater thickness. Figure 8.2a shows magnetoresistance for rotated magnetic field, from perpendicular 90°, to in plane current aligned, 0° orientation on a 14-layer sample. Negative magnetoresistance, or positive magnetoconductance is also prominent on this sample, having a $B^{1/2}$, which was also supressed by surface doping as shown in Fig. 8.2b, also confirming its bulk origins.

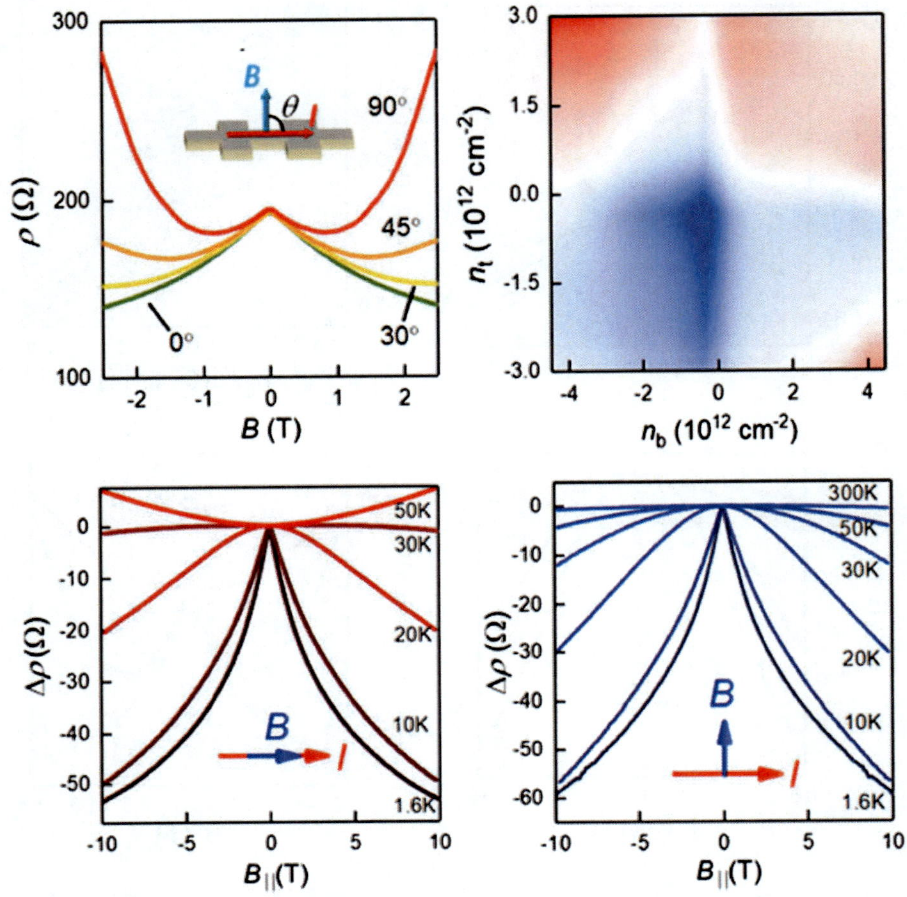

Fig. 8.2 Weak localization crossover and unidirectional in plane field negative magnetoresistance **a** weak field magneto-resistivity for perpendicular (90°) to in plane (0°) orientations of field. **b** Top–bottom surface carrier density dependence of magnetoresistance at B = 10 T, T = 1.6 K with colour scale blue to red −40 to 40 Ω. **c, d** Temperature dependence of magnetoresistance for current aligned and perpendicular in plane magnetic fields, respectively

Figure 8.2c, d shows temperature dependence of negative magnetoresistance for current aligned, and perpendicular in plane magnetic fields. In both cases, $B^{1/2}$ negative magnetoresistance below 20 K recovers to a quadratic dependence at higher temperature. While becoming positive at 50 K for current aligned magnetic field, the negative magnetoresistance persists until room temperature for perpendicular in plane magnetic field. The current aligned magnetoconductance at 1.6 K was found to exhibit a tendency to saturation at ±18 T as shown in Fig. 8.3 with magnetoresistance reaching just over a 100%, surprisingly high for a non-magnetic system, being comparable to the ones measured from disordered bulk states of more usual topological insulator systems.

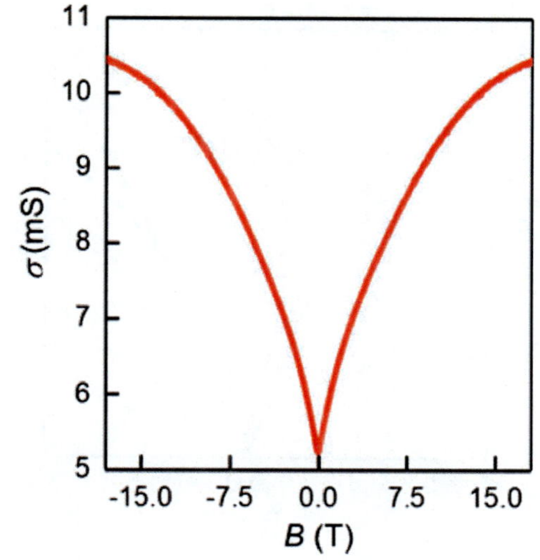

Fig. 8.3 Magnetoconductivity for a current aligned magnetic field of up to 18 T, with magnetoconductance approaching above 100% at maximum field

We believe the most likely reason behind the observed quadratic negative magnetoresistance resembling chiral anomaly is magnetic field induced suppression of scattering from Coulomb impurities, which arise on defects within the bulk of the rhombohedral graphite film and is independent of the magnetic field orientation. We note that, the uniform contacts across the entire width of the sample studied rules out any possible current jetting effects and such a negative magnetoresistance has been found to be absent on reference hexagonal stacked thin films.

We find that, the crossover into negative square root of magnetic field negative magnetoresistance on 14 layer device to be well described by the three-dimensional weak localization model [5]

$$\Delta\sigma = \frac{\alpha e^2}{2h}\left[\frac{\Psi\left(l_B^2/l_e^2\right)}{l_e} - \frac{\Psi\left(l_B^2/l_\varphi^2 + 1\right)}{l_\varphi} - \int_{1/l_\varphi}^{1/l_e}\Psi\left(l_B^2x^2 + \frac{1}{2}\right)dx - 2\left(\frac{1}{l_e} - \frac{1}{l_\varphi}\right)\right] \tag{8.3}$$

as shown in Fig. 8.4 (dashed dotted lines) with an additional, defects related βB^2 term as in the case of thinner samples. The term α is a system related coefficient, with $\Psi(z)$ being a digamma function, with characteristic lengths l_e, l_B and l_φ being mean free path, magnetic length and dephasing length respectively. The dephasing length l_φ was found to be ~0.1 μm being order of magnitude higher than calculated mean free path of 19 nm, from an extracted bulk electron mobility of 1040 cm^2/Vs (Chap. 4, Sect. 4.3) and an estimated lowest bulk conduction band effective mass of 0.0279 m_e. We attribute the crossover into weak localization on thicker samples to enhanced thermally activated bulk carrier density, and the decreasing intervalley

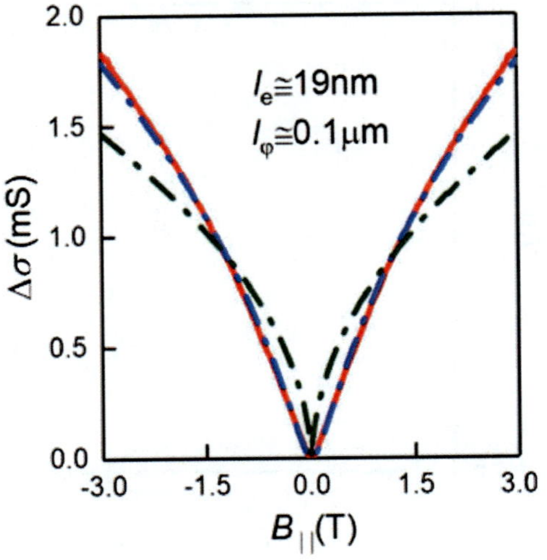

Fig. 8.4 Weak field magnetoconductance at 1.6 K (red line) fitted with Eq. (8.3) for dephasing lengths of 0.1 μm (dashed dotted blue line) and 1 μm (dashed dotted green line)

scattering length arising from it [12]. The dephasing length of 0.1 μm is an order of magnitude lower than the one that has been found for two dimensional Dirac fermions of graphene systems for which quantum interference effects were found to persist until up to an order of magnitude higher temperatures, ~200 K [12].

The observed uni-directional field dependence of negative magnetoresistance for 14-layer system is a strong proof for the argument of three-dimensional quantum interference. We believe the observed anisotropy shown in Fig. 8.5a, which is the

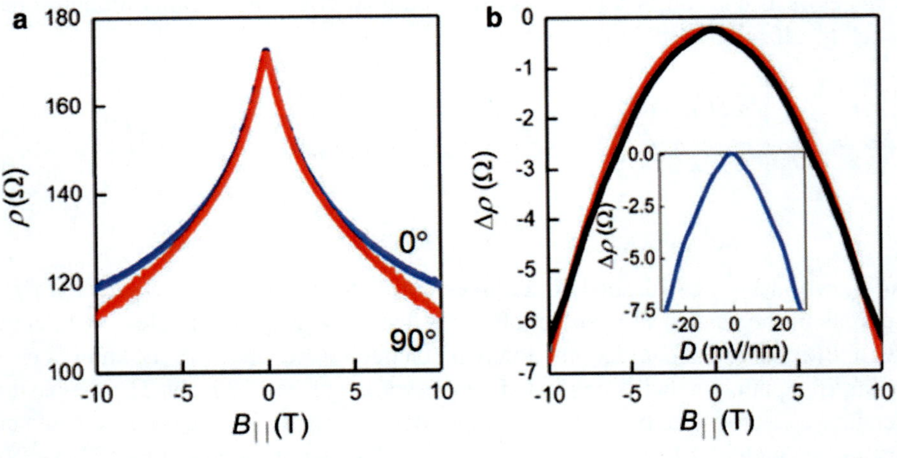

Fig. 8.5 **a** Anisotropic negative magnetoresistance of current aligned (red) and perpendicular (blue) in plane magnetic fields. **b** The difference in magnetoresistance (black) at 1.6 K and a quadratic fit (red) with the inset showing weak displacement field induced negative resistance around the charge neutrality point

additional quadratic component of magnetoresistance as in plane field is rotated to be perpendicular to the current, is due to an effective displacement field generation as pointed out in a recent work [13]. Such a Lorentz force induced layer polarisation would indeed have a quadratic dependence to magnetic field as shown in Fig. 8.5b (red fit). A negative resistance induced by small displacement field at the charge neutrality point of a 14-layer system at the charge neutrality point is shown on the inset of Fig. 8.5b, which suggests the generated effective displacement field by a field of ± 10 T to be $\cong 25$ mV/nm.

8.4 Summary

In summary, we have demonstrated three-dimensional quantum interference effects influenced by Berry phase and intervalley scattering on carbon-based nodal line semimetal thin films. We have found that as well as leading to a quadratic negative magnetoresistance resembling chiral anomaly, which persists until room temperature, bulk states undergo three dimensional diffusive motion, leading to a Berry phase induced weak antilocalization below 20 K, thus reflecting the topological properties of the bulk bands. For increasing thicknesses of 14 layers and beyond, as it has been found in graphene, a crossover into weak localization occurs due to a decreasing intervalley scattering length which becomes comparable to the dephasing length of 0.1 μm. Furthermore, we have shown that, in plane magnetic field when perpendicular to the current applied, generates a small displacement field, yielding an additional negative magnetoresistance contribution. Overall, we promote rhombohedral graphite to be an exciting platform not just as an N-layer generalisation of graphene in the low energy limit, but also due to three-dimensional transport its bulk charges display, strongly influenced by the bulk topology. We have demonstrated quantum interference effects seen in two dimensions in graphene, in three-dimensions despite the thin film nature of the rhombohedral graphite systems studied.

References

1. Koshino M (2010) Interlayer screening effect in graphene multilayers with ABA and ABC stacking. Phys Rev B Condens Matter Mater Phys 81(12):1–7. https://doi.org/10.1103/PhysRevB.81.125304
2. Burkov AA, Hook MD, Balents L (2011) Topological nodal semimetals. Phys Rev B Condens Matter Mater Phys 84(23):1–14. https://doi.org/10.1103/PhysRevB.84.235126
3. Bian G et al (2016) Topological nodal-line fermions in spin-orbit metal PbTaSe2. Nat Commun 7:1–8. https://doi.org/10.1038/ncomms10556
4. Ali MN et al (2016) Butterfly magnetoresistance, quasi-2D Dirac fermi surface and topological phase transition in ZrSiS. Sci Adv 2(12):1–8. https://doi.org/10.1126/sciadv.1601742
5. Chen W, Lu HZ, Zilberberg O (2019) Weak localization and antilocalization in nodal-line semimetals: dimensionality and topological effects. Phys Rev Lett 122(19):196603. https://doi.org/10.1103/PhysRevLett.122.196603

6. Kim HJ et al (2013) Dirac versus weyl fermions in topological insulators: Adler-Bell-Jackiw anomaly in transport phenomena. Phys Rev Lett 111(24):1–5. https://doi.org/10.1103/PhysRevLett.111.246603
7. Huang X et al (2015) Observation of the chiral-anomaly-induced negative magnetoresistance: in 3D Weyl semimetal TaAs. Phys Rev X 5(3):1–9. https://doi.org/10.1103/PhysRevX.5.031023
8. Li Q et al (2016) Chiral magnetic effect in ZrTe5. Nat Phys 12(6):550–554. https://doi.org/10.1038/nphys3648
9. Zhang CL et al (2016) Signatures of the Adler-Bell-Jackiw chiral anomaly in a Weyl fermion semimetal. Nat Commun 7:1–9. https://doi.org/10.1038/ncomms10735
10. Kawabata A (1980) Theory of negative magnetoresistance in three-dimensional systems. Solid State Commun 34(6):431–432. https://doi.org/10.1016/0038-1098(80)90644-4
11. Altshuler BL, Khmel'nitzkii D, Larkin AI, Lee PA (1980) Magnetoresistance and Hall effect in a disordered two-dimensional electron gas. Phys Rev B 22(11):5142–5153. https://doi.org/10.1103/PhysRevB.22.5142
12. Tikhonenko FV, Kozikov AA, Savchenko AK, Gorbachev RV (2009) Transition between electron localization and antilocalization in graphene. Phys Rev Lett 103(22):1–4. https://doi.org/10.1103/PhysRevLett.103.226801
13. Slizovskiy S, McCann E, Koshino M, Fal'ko VI (2019) Films of rhombohedral graphite as two-dimensional topological semimetals. Commun Phys 2(1):1–10. https://doi.org/10.1038/s42005-019-0268-8

Chapter 9
Summary

Rhombohedral graphite films as topological insulators exhibit an interplay of intrinsic semiconductor-like thermally activated bulk conduction and a low temperature surface dominated conduction.

Landau level spectroscopy of the surfaces at low temperatures revealed that $E \sim p^N$ dispersion of flat bands predicted by the simplest tight binding model is not representative of the low energy band structure of hBN encapsulated rhombohedral graphite devices studied. Rather, a trigonal warping leading to 3-degenerate valance band maxima is found, with a $E \sim p^2$ dispersion in low momentum limit. Albeit $E \sim p^N$ dispersion picture not being validated, the low energy bands are still flat, with a Berry phase of $N\pi$ and are still susceptible to a low temperature, BCS-like superconductivity.

The $E \sim p^2$ dispersing flat bands exhibit spontaneous inversion symmetry breaking below 15 K for the case of 9–12 layers thick films, accompanied by topological currents arising due to a generated Berry curvature. I–V characterisation at the charge neutrality point yields an unusual negative differential resistance (NDR). This NDR onset voltage was studied both as a function of temperature and magnetic field yielding fits to mean field expressions as the gap is destroyed as a function of each parameter. Moreover, in-plane magnetic field dependence suggested a gap destruction due to Zeeman energy enhancement, with the NDR onset voltage decreasing linearly as a function of in-plane magnetic field. Hysteretic behaviour of gapped state was found, as a function of magnetic field and doping, suggesting presence of multiple mesoscopic domains with a finite magnetisation.

A gap opening was also found as a function of applied displacement field, which was previously theoretically predicted. This gap was found to be absent in both theoretical calculations and experiments on the identified graphite regions with stacking faults.

Lastly, transport properties of bulk charges were also investigated by carrying out magneto-transport experiments with varying orientations of magnetic field. A negative magnetoresistance with a quadratic magnetic field dependence, resembling chiral anomaly was found for all orientations of in plane magnetic field. Below

S. Ozdemir, *Electronic Properties of Rhombohedral Graphite*, Springer Theses,
https://doi.org/10.1007/978-3-030-88307-2_9

20 K however, this negative magnetoresistance which we attribute to a magnetic field induced suppression of scattering from Coulomb impurities was overwhelmed by weak anti-localization for the case of smaller number of bulk electron density, reflecting the topologically non-trivial nature of gapped bulk Bloch bands. With decreasing bulk band gap, and hence with increasing bulk carrier density the quantum interference below 20 K was found to crossover into weak localization, being perfectly well described by a 3D weak localization model and yielding a dephasing length on the order of 0.1 μm. This explains the order of magnitude lower temperatures necessary for observation of the effects in contrast to 200 K for dephasing length on the order of 1 μm on high quality graphene devices. Lastly, when the orientation of in plane magnetic field was set perpendicular to current, a Lorentz force induced effective displacement field generation was found, also with a quadratic dependence on applied magnetic field.

As future work, there are yet experiments to do to confirm whether the predicted superconductivity exists in flat bands of rhombohedral graphite. We have seen some features in I–V curves that may suggest a superconducting phase in the vicinity of the gapped state away from the NDR region with the experiments possibly being subject of another a Ph.D. thesis. Aside from the superconductivity experiments, low temperature scanning tunnelling spectroscopy experiments could be carried out in order to physically map the bulk-surface transition taking place at low temperatures, as well as the mesoscopic domains that arise due to spontaneous symmetry breaking at the flat bands. Finally, due to Berry phase of $N\pi$ the low energy flat bands possess, one expects an odd/even layer weak localization/antilocalization in transport experiments if we were to exclude the bulk charges completely, which is possible at extremely low temperatures around 10 mK. One can also take the growth experiments carried out on SiC substrates further to see if it is possible to grow rhombohedral graphite films beyond 5-layers and characterise it in a comparative study with exfoliated rhombohedral graphite domains.

Overall, we have reported the first ever systematic study of rhombohedral graphite thin films of up to 50 layers, having characterised its electronic structure through electron transport, we have shown that similar to twisted bilayer magic angle systems, it is home to and exhibits correlated system physics that is typically present on more traditional carefully grown heavy fermion system crystals. Major advantage of our system in comparison to twisted bilayer graphene is that it does not require careful aligning of two crystals, which is currently a major fabrication challenge. As well as the two-dimensional transport, bulk of rhombohedral graphite films exhibit three-dimensional quantum interference influenced by bulk topology, which has so far been seen on exotic crystal systems such as Weyl and Dirac semimetals.

Curriculum Vitae

Servet Ozdemir

Post-Doctoral Research Associate | Quantum Photonics Laboratory, Institute of Photonic and Quantum Sciences, Heriot-Watt University

Email: s.ozdemir@hw.ac.uk

I am currently a post-doctoral research associate working at Heriot-Watt University under Prof. Brian Gerardot. I am currently making HWU's first electron transport devices on van der Waals heterostructures of 2D crystals as well as establishing electron transport set-ups in the Quantum Photonics Laboratory. During my Ph.D. I was working under Sir Konstantin Novoselov and Prof. Artem Mishchenko at the University of Manchester where I have carried out first ever systematic investigation of electron transport studies of rhombohedral graphite films. I was also involved in various projects like tunnelling spectroscopy of novel hBN embedded graphene quantum dots and 2D magnetism.

Employment History

09/2020–Present

Post-Doctoral Research Associate

- Experiments on simultaneous optical and electrical characterisation of van der Waals heterostructures including sample fabrication.

07/2016—Present

PhD Student | The University of Manchester

- PhD on electronic properties of van der Waals heterostructures of 2D crystals.

S. Ozdemir, *Electronic Properties of Rhombohedral Graphite*, Springer Theses,
https://doi.org/10.1007/978-3-030-88307-2

10/2015–06/2016

MPhys Student | University of Warwick

- Low temperature electronic transport properties of stained Germanium quantum wells.

07/2015–09/2015

Summer Research Assistant | University of Cambridge

- Nanoscale electronic device fabrication on electro-deposited nanowires under supervision of Dr. Liam O'Brien in the group of Prof. Russel Cowburn FRS.

07/2013–06/2015

Visiting Undergraduate Research Student | Bogaziçi University

- Modelling and data analysis of a resistance drift due to a structural relaxation in amorphous phase change nano-memory devices under supervision of Prof. Özhan Özatay.

Education

09/2016–09/2020

The University of Manchester

PhD: Mesoscopic Physics

Supervisor: Sir Konstantin Novoselov, Prof. Artem Mishchenko.

Thesis Title: [Electronic properties of rhombohedral graphite].

10/2012–09/2016

University of Warwick

Master of Physics, Bachelor of Science

Thesis Title: Anisotropy in low temperature electron transport properties of strained high mobility Germanium quantum wells.

Publications

Journals:

Electronic phase separation in multilayer rhombohedral graphite, Yanmeng Shi et al., Nature, Vol. 584, August 2020.

Stacking order in graphite films controlled by van der Waals technology, Yaping Yang et al., Nano Letters, Vol. 19, October 2019.

Dimensional reduction,quantum Hall effect and layer parity in graphite films, Jun Yin et al., Nature Physics, Vol. 15, May 2019.

Stacking transition in rhombohedral graphite, Tataiana Latychevskaia et al., Frontiers of Physics, Vol. 14, February 2019.

Planar and van der Waals heterostructures for vertical tunnelling single electron transistors, Gwangwoo Kim et al., Nature Communications, Vol 10, January 2019.

Graphene hot-electron light bulb: incandescence from hBN-encapsulated graphene in air, Seok-Kyun Son et al., 2D Materials, Vol. 5., November 2017.

Toward multiple-bit-per-cell memory operation with stable resistance levels in phase change nanodevices, Aisha Gokce et al., IEEE Transactions on Electron Devices, Vol. 63, Aug 2016.

Book Chapters:

Novel Phenomena in 2D Semiconductors, Servet Ozdemir et al.,**2D Semiconductor Materials and Devices**, Elsevier, November 2019.

Talks and Presentations

12/06/2017—Poster presentation "**Coloumb Diamonds in hBN converted Graphene Quantum Islands**" on CDT Student conference on Graphene Technology, Wyboston Lakes, UK.

02/07/2018—Poster presentation "**Graphene hot-electron light bulb: incandescence from hBN-encapsulated graphene in air**" on CDT Student conference on Graphene and 2D Materials, DeVere Cranage Estate, UK.

18/06/2019—Talk "**Quantum transport on topologically protected flat bands of ABC stacked multilayer graphene**" on Quantum Transport one-day Workshop, University of Warwick.

02/07/2019—Talk "**Quantum transport on topological insulator thin films of rhombohedral graphite**" on CDT Student conference on Graphene and 2D Materials, Wyboston Lakes, UK.

30/11/2019—Keynote speaker talk "**Quantum transport on topologically protected flat bands of ABC stacked multilayer graphene**" on ICNAN-19, India.

02/12/2019—Invited talk "**Tunnelling spectroscopy of 2D Anti-ferromagnetism on a-RuCl3 crystals**" IIT Madras, Department of Physics, India.

02/12/2019—Invited talk "**Quantum transport on topological insulator thin films of rhombohedral graphite**" Bogazici University, Department of Physics, Turkey.

Awards

Graphene NOWNANO Centre of Doctoral Training PhD Studentship—Sep. 2016–Sep. 2020.